T0135479

Aspects of regularization in Banach spaces

Kamil S. Kazimierski

Bibliografische Information der Deutschen Nationalbibliothek

Die Deutsche Nationalbibliothek verzeichnet diese Publikation in der
Deutschen Nationalbibliografie; detaillierte bibliografische Daten sind
im Internet über http://dnb.d-nb.de abrufbar.

ISBN 978-3-8325-2731-0

Logos Verlag Berlin GmbH
Comeniushof, Gubener Str. 47,
10243 Berlin
Tel.: +49 (0)30 42 85 10 90
Fax: +49 (0)30 42 85 10 92
INTERNET: http://www.logos-verlag.de

Aspects of regularization in Banach spaces

von Kamil S. Kazimierski

Dissertation

zur Erlangung des Grades eines Doktors der Naturwissenschaften
— Dr. rer. nat. —

Vorgelegt im Fachbereich 3 (Mathematik & Informatik)
der Universität Bremen
im April 2010

Datum des Promotionskolloquiums: 30. Juni 2010

Gutachter: Prof. Dr. Peter Maaß (Universität Bremen)
Prof. Dr. Thomas Schuster (Universität Hamburg)

Zusammenfassung

In letzten Jahren gab es ein wachsendes Interesse an schlecht-gestellten inversen Problemen für Operatoren, welche zwischen zwei Banach-Räumen abbilden. In dieser Dissertation werden wir ein besonderes Augenmerk legen auf den Fall von linearen, stetigen Operatoren und Banach-Räumen welche konvex vom Potenztyp und/oder glatt vom Potenztyp sind. Wir werden sowohl die Tikhonov-Regularisierung wie auch die Landweber-Regularisierung betrachten.

Es ist bekannt, dass viele der üblichen Räume, wie die Folgenräume ℓ^p, Lebesgue-Räume $L^p(\Omega)$ und Sobolev-Räume $W_k^p(\Omega)$ zusammen mit den üblichen Normen konvex vom Potenztyp und glatt vom Potenztyp sind, wenn $1 < p < \infty$. In dieser Dissertation werden wir zeigen, dass für $1 < p, q < \infty$ auch die Wavelet-Charakterisierung der Norm auf Besov-Räumen $B_{p,q}^s(\mathbb{R})$ konvex vom Potenztyp und glatt vom Potenztyp ist.

Der zweite Teil dieser Dissertation beschäftigt sich mit der Tikhonov-Regularisierung. Einige Begrifflichkeiten, welche in der Regularisierung in Hilbert-Räumen verwendet werden, wie Quellbedingungen, a-priori Parameterwahl, Diskrepanzprinzip von Morozov, wurden bereits in das Setting der Banach-Räume übersetzt. Wir werden zwei neue Beiträge machen. Erstens werden wir das Diskrepanzprinzip von Engl für das Setting der Banach-Räume verallgemeinern und zeigen, dass, unter geeigneten Quellbedingungen, bessere Konvergenzraten als die für das Diskrepanzprinzip von Morozov bekannten Raten erreicht werden können. Zweitens werden wir Konvergenzraten für zwei Minimierungsmethoden des Tikhonov-Funktionals zeigen.

Im dritten Teil werden wir die Landweber-Regularisierung betrachten, für die wir ebenfalls zwei neue Beiträge liefern werden. Erstens werden wir eine für Banach-Räume, welche konvex vom Potenztyp und glatt vom Potenztyp sind, angepasste Version der Landweber-Iteration von [59] vorstellen und starke Konvergenz zeigen. Zweitens werden wir eine modifizierte Version der Landweber-Iteration einführen, für die wir unter einer geeigneten Quellbedingung eine Konvergenzrate zeigen werden.

Zum Schluss werden wir numerische Beispiele für die in dieser Dissertation besprochenen Algorithmen vorstellen.

Abstract

In recent years there has been an increasing interest in the regularization of ill-posed inverse problems for operators mapping between two Banach spaces. In this thesis we will focus on the case of linear, continuous operators and Banach spaces, which are convex of power type and/or smooth of power type. We will consider the Tikhonov regularization as well as the Landweber regularization.

It is well-known that many common spaces, like the sequence spaces ℓ^p, Lebesgue spaces $L^p(\Omega)$ and Sobolev spaces $W_k^p(\Omega)$ equipped with the usual norms are smooth of power type and convex of power type if $1 < p < \infty$. In this thesis we will show that the wavelet characterization of the norm of Besov spaces $B^s_{p,q}(\mathbb{R})$ is convex of power type and smooth of power type for $1 < p, q < \infty$.

A second major part of this thesis deals with Tikhonov regularization. Some notions used in regularization in Hilbert spaces like source conditions, a-priori parameter choice, Morozov's discrepancy principle have already been translated to the setting of Banach spaces. We will make two new contributions. First, we will generalize the discrepancy principle of Engl to the setting of Banach spaces and show that under appropriate source conditions better convergence rates than those known for Morozov's discrepancy principle may be achieved. Second, we will show convergence rates for two minimization methods for the Tikhonov functional.

In the third part, we will consider Landweber regularization, where two new contributions will be made. First, we will adapt the Landweber iteration of [59] to Banach spaces convex of power type and smooth of power type. Further, strong convergence will be shown. It will be followed by the introduction of a modified version of the Landweber iteration. Finally, we will develop a convergence rate under an appropriate source condition.

The quality of the algorithms introduced in this thesis will be discussed with help of several numerical examples.

Tell me everything you know.
I want to know everything.
I want to know.

Col. Dr. IRINA SPALKO

I would like to thank:

Prof. Dr. Peter Maaß for supervision, support and motivation in writing this thesis. All the people from the *Zentrum für Technomathematik* for mathematical and technical support, especially the whole *AG Technomathematik* for an excellent working atmoshpere and the nice time I had and even more special Dirk Hentschel, Christina Brandt, Dr. Kristian Bredies, Dr. Torsten Hein, Dr. Dorota Kubalińska, Dr. Iwona Piotrowska, Stefan Schiffler, Rudolf Ressel and Dr. Frank Schöpfer for support, advices and discussions - both mathematically and personally.

Kamil S. Kazimierski
Zentrum für Technomathematik
Universität Bremen

Contents

1

Introduction

This dissertation is devoted to the regularization of linear, ill-posed problems with methods employing Banach space norms, which are convex of power type and/or smooth of power type. Such regularization methods make use of techniques and methods coming from many different areas of mathematics: geometry of Banach spaces, optimization, classical regularization theory, functional analysis just to name a few.

We deal with linear, continuous operators A mapping between two Banach spaces X and Y. Our aim is to find a good approximation of the solution of the operator equation

$$Ax = y,$$

where only a noise cluttered version y^δ of the true data y and the noise level δ with

$$\|y - y^\delta\| \leq \delta$$

are known. This problem is not trivial if the naïve approach of substituting x by x^δ with

$$Ax^\delta = y^\delta$$

fails. This may be the case if the operator A is not surjective on the space of all possible cluttered versions of the data; if the operator A is not injective or if the formal inverse A^{-1} is not continuous. Such problems are called ill-posed and the method of finding approximate solutions of $Ax = y$ is called regularization.

Typically, ill-posed problems appear in mathematical models of measurement processes in physics, medicine, biology, meteorology, seismology, photography, econometrics and many other fields of science.

This thesis makes contributions to the case, where the underlying Banach spaces X and Y are convex of power type and/or smooth of power type. We show new results and their connection to other previously known results.

In the next section, we present an example of an ill-posed problem. Namely, we discuss the retrieval of the velocity information from GPS data. Further, we show how the use of Banach space norms arises naturally in our setting.

In the next chapter, we introduce all basic notions which are necessary in the study of regularization methods, which employ Banach space norms. First, we present a generalization of the derivative for convex functions known as the *subgradient*. In particular the so called *duality mapping*, which is the subgradient of a power of a Banach space norm. Next, we define the notion of a Banach norm *smooth of power type* and *convex of power type*. Further, we introduce the notion of *Bregman distances* and discuss their connection to the norm. Finally, we introduce the notion of *minimum norm solution*, which is essential in the study of properties of a regularization scheme.

It is well-known that the sequence spaces ℓ^p, the Lebesgue spaces $L^p(\Omega)$, the Sobolev spaces $W_m^p(\Omega)$ are smooth of power type and convex of power type for $1 < p < \infty$. Aside from the Sobolev scale, the scale of Besov spaces $B_{p,q}^s(\mathbb{R})$ is being widely used to describe the smoothness of a function. The Besov scale is especially popular in image processing, where the norm on the Besov spaces is usually characterized via coefficients of the wavelet expansion of the underlying function. The main aim of Chapter 3 is to show that this characterization of the norm is smooth of power type and convex of power if $1 < p, q < \infty$.

In Chapter 4, we consider Tikhonov regularization with Banach space norms smooth of power type and convex of power type. First, we restate known results, which ensure that under some mild conditions on the pre-image space the minimizers of the Tikhonov functional

$$T_\alpha(x) := \frac{1}{p}\|Ax - y^\delta\|^p + \alpha\frac{1}{q}\|x\|^q$$

converge to the minimum norm solution. The next goal of the chapter is to show convergence rates with respect to Bregman distance of the minimizers of the Tikhonov functional to the minimum norm solution. From classical regularization theory it is well-known that this is only possible if additional assumptions on the minimum norm solution — the so-called source conditions — are made and if the parameter α is chosen appropriately. Therefore, the next part of the chapter deals with the generalization of this notions into the Banach space setting. In particular, we introduce a generalization of the Engl's discrepancy principle as an a-posteriori parameter choice rule. It is followed by a study of two minimization algorithms of the Tikhonov functionals.

Chapter 5 deals with the Landweber regularization, which is an iterative regularization scheme and may be regarded as a steepest descent algorithm for the functional $x \mapsto \frac{1}{p}\|Ax - y^\delta\|^p$. We study a version of the Landweber iteration of [59], which we alter to fit best our setting. We also introduce a modified version of the Landweber iteration for which we show convergence rates under an appropriate source condition.

In Chapter 6, we present numerical experiments for some of the algorithms developed in this thesis.

We conclude this thesis with some final remarks and the list of new results.

1.1 Example: Velocity retrieval from GPS data

In this section, we present the retrieval of velocity from GPS data as an example of a linear, ill-posed problem. Moreover, we motivate the use of Banach space norms in the regularization of this problem.

We start with the following question: Given the GPS coordinates of a vehicle, is it possible to recover its velocity? We know that for given velocity v the position of the vehicle may be computed via

$$s(T) = \int_0^T v(t)\, dt + s(0).$$

We can choose the coordinates such that $s(0) = 0$. We know[1] that the Global Positioning System (GPS) has currently a guaranteed accuracy of about 7 meters. Hence, we can bound the distance between the exact position s and the GPS position s^δ by some number δ.

Altogether we have the setting

$$As = v, \qquad \|s - s^\delta\| \leq \delta,$$

where A is the integral operator $(As)(T) = \int_0^T s(t)\, dt$ defined on a pre-image space big enough to contain all reasonable velocities and mapping to an image space big enough to contain all noisy versions s^δ of the exact position s. We also notice that the pre-image and the image space shall be chosen in such a way that the operator is continuous. We assume that the GPS data is available only for a finite interval of time, say 1 hour. Hence, in our case, the choice $X = L^2(0, 1)$ for the pre-image space and $Y = L^2(0, 1)$ for the image space is sensible.

In Figure 1.1, we depict an example of the true velocity, the resulting exact position and the simulated GPS position. We see that the noice clutter is neglible compared to the value of the position ($7\,m$ compared to about $50\,km$). We know that $As = v$ and we observe that s and s^δ are hardly distinguishable. Hence, in a first, naïve approach it seems to be reasonable to approximate v by v^δ given by

$$Av^\delta = s^\delta.$$

We assume, for the sake of argument, that the above equation is solvable in some appropriate sense. In Figure 1.1, one can see an example of such naïve reconstruction. Obviously, the naïve reconstruction has nothing in common with the true velocity.

One can see that the naïve reconstruction v^δ is having much more oscillations than the true velocity v. Therefore, the norm of v^δ is much bigger than the norm of v. Hence, we implicitly assume that the true solution has a small norm. However, this additional information was not used in the reconstruction process.

---

[1] http://users.erols.com/dlwilson/gpsacc.htm

We improve our reconstruction approach by incorporating the information about the true solution. On the one hand, we want our improved reconstruction v_α^δ to solve $Av_\alpha^\delta = s^\delta$, at least approximately. Hence, we want v_α^δ to be such that $\frac{1}{2}\|Av_\alpha^\delta - y^\delta\|_2^2$ is small. On the other hand, we want v_α^δ to be such that the number $\frac{1}{2}\|v_\alpha^\delta\|_2^2$ is small too. Hence, in order to fulfill both requirements at the same time we may choose v_α^δ as

$$v_\alpha^\delta = \mathrm{argmin}_v \, \tfrac{1}{2}\|Av - y^\delta\|_2^2 + \alpha\tfrac{1}{2}\|v\|_2^2.$$

The parameter α allows us to control the balance between the first and the second term in the right-hand side of the above equation. In a sense, it allows us to control the meaning of the words *small* and *big*. The functional $v \mapsto \frac{1}{2}\|Av - y^\delta\|_2^2 + \alpha\frac{1}{2}\|v\|_2^2$ is called *Tikhonov functional*. As one can see in Figure 1.1, the Tikhonov approach significantly improves the reconstruction, if the parameter α is chosen appropriately.

Next, we will explain why in some situations it is advisable to use other norms than the L^2-norm in order to describe the true solution. To do so we consider the velocity profile depicted in Figure 1.2, which consists of three sharp peaks. As we can see, in Figure 1.2, the Tikhonov reconstruction proposed above is not able to reconstruct simultaneously the height and the support of the peaks. However, this time we implicitly know that the true solution has a small support. This is a different kind of information than we had before. Therefore, instead of using the L^2-norm we need to use a functional which penalizes small entries in the reconstruction more than the L^2-norm. A family of natural candidates are the L^p-norms with $1 < p \leq 2$. Due to bad geometrical properties of the L^1-norm, we will not consider the case $p = 1$ in this thesis. Then, the related Tikhonov functional may be defined via

$$\tfrac{1}{2}\|Av - y^\delta\|_2^2 + \alpha\tfrac{1}{p}\|v\|_p^p.$$

This approach improves significantly the quality of the reconstruction, as we can see in Figure 1.2

We may summarize the properties of the above example: The measurement setting is modeled via a linear, continuous operator mapping between two normed spaces. Usually, both spaces are infinite dimensional. The pre-image space is chosen such that it contains the set of all sensible sources of the measurements. Further, the image space is chosen such that it contains not only all resulting measurement data but also all expected perturbed or noise cluttered versions of the measurements. Moreover, the information about the magnitude of the perturbation of the measurement is available.

Further, we imply that some knowledge about the set of all reasonable solutions is available. We also assume that the information about that set may be encoded via an appropriate Banach norm.

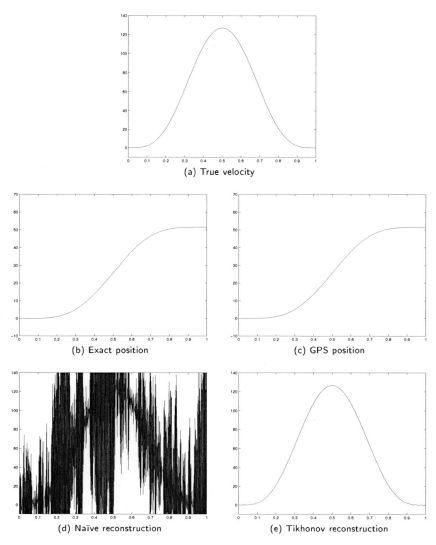

Figure 1.1: Naïve and Tikhonov reconstruction of the velocity from GPS coordinates (x-axis: time in hours; y-axis: velocity in km/h resp. position in km).

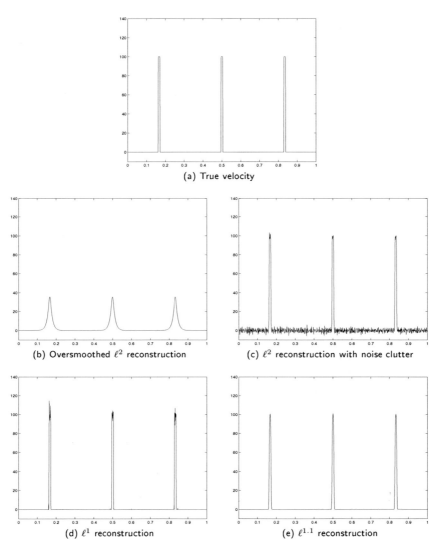

Figure 1.2: True velocity and Tikhonov reconstructions from GPS coordinates for different penalties based on Hilbert and Banach space norms (x-axis: time in hours; y-axis: velocity in km/h).

2

Preliminaries and basic definitions

In this chapter, we will introduce the definitions of concepts which are essential for this work. We will start with general nomenclature, introduce the main definitions from convex analysis and proceed with definitions and facts from geometry of Banach spaces. Finally, in the last part of this chapter we will discuss some notions from theory of regularization.

The main aim of the convex analysis part is to introduce the notion of subgradients, which can be thought of as a generalization of the notion of gradients to convex (non-smooth) functions. A very special example will be the subgradient of a power of the norm. We will call such a mapping duality mapping. This mapping will be essential in all analysis used in this work. We will present the most interesting properties of the duality mapping. The most prominent among them is the Theorem of Asplund which states that the duality mapping is in fact the subgradient of a power of the norm. Further, the duality mapping may be used to transport elements from the primal space into the dual space and vice versa.

The part on the geometry of Banach spaces has two aims. The first one is to introduce the notion of smoothness and convexity of power type, which are central throughout this thesis. The second aim is the introduction of Bregman distances, which turn out to be to better suited for convergence analysis than the norm. As we will see, for spaces smooth of power type resp. convex of power type there is a tight connection between the Bregman distance and the norm.

2.1 General

We start with some standard definitions, inequalities and nomenclature. A *generic constant*

$$C > 0$$

can take different value every time it is used. We use this notation in some proofs to avoid unnecessary numeration of constants.

Definition 2.1 (Conjugate exponents). *For $p \geq 1$ we denote by $p^* \geq 1$ the conjugate exponent of p given by*

$$\frac{1}{p} + \frac{1}{p^*} = 1.$$

Definition 2.2 (Dual space). *We denote by X^* the (topological) dual space of X, i.e. the space of all continuous linear forms mapping from X to $\mathbb{R}$ equipped with the norm*

$$\|x^*\|_{X^*} := \sup_{\|x\|=1} |x^*(x)|.$$

Definition 2.3. *Fox $x^* \in X^*$ and $x \in X$ we denote by $\langle x^*, x \rangle_{X^* \times X}$ and $\langle x, x^* \rangle_{X \times X^*}$ the duality pairing (dual mapping or dual composition) defined via*

$$\langle x^*, x \rangle_{X^* \times X} := \langle x, x^* \rangle_{X \times X^*} := x^*(x).$$

We will omit the indices indicating the spaces where they are clear from the context. Further, we emphasize that the duality pairing is not to be confused with the duality mapping, which will be introduced in Definition 2.24.

Theorem 2.4 (Adjoint operator). *Let X, Y be normed spaces and $A : X \rightarrow Y$ be continuous and linear. Then, the continuous operator $A^* : Y^* \rightarrow X^*$ defined via*

$$\langle Ax, y^* \rangle_{Y \times Y^*} = \langle x, A^* y^* \rangle_{X \times X^*} \qquad \forall x \in X, y^* \in Y^*$$

is called the adjoint operator (for the proofs of well-definition and continuity cf. [51, Section 3.1]).

Definition 2.5. *By $\mathcal{N}(A)$ we denote the null-space of the operator A, i.e. $\mathcal{N}(A) := \{x : Ax = 0\}$. By $\mathcal{R}(A)$ we denote the range of A, i.e. $\mathcal{R}(A) := \{y : \exists x : y = Ax\} = A(X)$.*

Theorem 2.6 (Cauchy's inequality). *Let $x \in X$ and $x^* \in X^*$ then*

$$|\langle x^*, x \rangle_{X^* \times X}| \leq \|x^*\|_{X^*} \cdot \|x\|_X .$$

Theorem 2.7 (Hölder's inequality). *Let $x \in \ell^p$ and $y \in \ell^{p^*}$ with $\frac{1}{p} + \frac{1}{p^*} = 1$ then*

$$\sum_k |x_k y_k| \leq \left(\sum_k |x_k|^p \right)^{1/p} \cdot \left(\sum_k |y_k|^{p^*} \right)^{1/p^*} .$$

Theorem 2.8 (Young's inequality). *Let $p, p^* > 1$ be conjugate exponents and $a, b \in \mathbb{R}$ then*

$$a \cdot b \leq \frac{1}{p} |a|^p + \frac{1}{p^*} |b|^{p^*}.$$

Definition 2.9 (Equivalent quantities). *We call two quantities a and b equivalent if there exist constants $0 < c, C < \infty$, such that*

$$c \cdot b \leq a \leq C \cdot b$$

and write in shorthand

$$a \sim b.$$

In particular for two norms $\| \cdot \|_A : X \to \mathbb{R}$ and $\| \cdot \|_B : X \to \mathbb{R}$ we mean by

$$\| \cdot \|_A \sim \| \cdot \|_B$$

that

$$c\|x\|_B \leq \|x\|_A \leq C\|x\|_B \qquad \forall x \in X,$$

for some constants $0 < c, C < \infty$, which do not depend on x.

If α is a function of δ then we mean by

$$\alpha \sim \delta^{p-1}$$

that

$$c \cdot \delta^{p-1} \leq \alpha(\delta) \leq C \cdot \delta^{p-1}$$

for some constants $0 < c, C < \infty$, which do not depend on δ.

Definition 2.10 (Gâteaux differentiability). *Let $f : X \to Y$ be a given map, with X and Y normed spaces. The map f is Gâteaux differentiable at x if there exists a continuous, linear map A_x from X to Y such that for all $h \in X$ we have*

$$\lim_{t \to 0} \frac{f(x + th) - f(x)}{t} = A_x h.$$

Definition 2.11 (Fréchet differentiability). *Let $f : X \to Y$ be a given map, with X and Y normed spaces. The map f is Fréchet differentiable at x if there exists a continuous, linear map A_x from X to Y such that for all $z \in X$ we have*

$$\lim_{z \to 0} \frac{\|f(x + z) - f(x) - A_x z\|}{\|z\|} = 0.$$

Remark 2.12. *If f is Fréchet differentiable at x, then it is also Gâteaux differentiable there. This can be seen by setting $z = th$ in the last definition.*

2.2 Convex analysis

Definition 2.13 (Convex functions). *The function $f : X \to \mathbb{R} \cup \{\infty\}$ is convex if*

$$f(\lambda x + (1 - \lambda)y) \leq \lambda f(x) + (1 - \lambda)f(y) \qquad \forall x, y \in X, \forall \lambda \in [0, 1].$$

Definition 2.14 (Effective domain, Proper functions). *Let* $f : X \to \mathbb{R} \cup \{\infty\}$ *be convex. The effective domain* $\operatorname{dom} f$ *of the function* f *is defined via*

$$\operatorname{dom} f := \{x \ : \ f(x) < \infty\}.$$

The function f *is* proper *if*

$$\operatorname{dom} f \neq \emptyset.$$

Example 2.15 (Tikhonov functional). *Let* X *and* Y *be Banach spaces,* $A : X \to Y$ *a linear, bounded operator,* $y^\delta \in Y$ *and* $p, q > 1$. *Then, for* $\alpha > 0$ *the functional* $\mathrm{T}_\alpha : X \to Y$

$$\mathrm{T}_\alpha(x) := \tfrac{1}{p} \|Ax - y^\delta\|_Y^p + \alpha \cdot \tfrac{1}{q} \|x\|_X^q \tag{2.1}$$

is called a Tikhonov functional. *Tikhonov functionals are convex and*

$$\operatorname{dom} \mathrm{T}_\alpha = X.$$

The rest of the section is driven by the question: What are the optimality conditions of the Tikhonov functional?

Remark 2.16. *Some authors use for the Tikhonov functional a symbol which emphasizes the dependence on the noisy data* y^δ, *say* $\mathrm{T}_{\alpha, y^\delta}$. *To keep the notation more compact we will use the notation introduced in Example 2.15. Where such compact notation would be confusing, we will write the functional in its full form.*

Definition 2.17 (Subgradient of convex functions). *Let* $f : X \to \mathbb{R} \cup \{\infty\}$ *be a convex function. Then,* $x^* \in X^*$ *is a* subgradient *of* f *at* x *if*

$$f(y) \geq f(x) + \langle x^*, y - x \rangle \qquad \forall y \in X.$$

The set

$$\partial f(x)$$

of all subgradients of f *at* x *is called the* subdifferential. *In what follows we will call both the subgradient and the subdifferential just subgradient, since it will be clear from the context which one is meant. Especially for single-valued subdifferentials we will identify the whole set with its single element.*

All affine functionals g which are tangential to f at x, i.e. $g(y) \leq f(y)$ for all y and $g(x) = f(x)$, can be written in the form

$$g(y) = \langle x^*, y - x \rangle + f(x).$$

Notice that the graph of g is a hyperplane. Therefore, geometrically, the subgradient is the slope (or set of slopes) of a hyperplane, which supports f (cf. Figure 2.1).

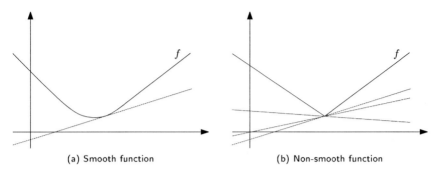

(a) Smooth function   (b) Non-smooth function

Figure 2.1: The subgradient of a function at x as the slope of the hyperplane supporting the function at x. For smooth functions the subgradient is single-valued, for non-smooth functions it is in general not single-valued.

We keep in mind that the main reason why the subgradient was introduced, was to get some information on the points for which the Tikhonov functional T_α is minimal. First, let us consider a simpler case of of a convex and differentiable function mapping from a one-dimensional space into another one-dimensional space, say $f : \mathbb{R} \to \mathbb{R}$. Then, the minimizer z of f is characterized by the optimality condition

$$f'(z) = 0.$$

The main purpose of the subgradient is the generalization of the above optimality condition to the case of non-smooth (convex) functions.

Theorem 2.18 (Optimality conditions). *[56, 4.1.2] Let $f : X \to \mathbb{R} \cup \{\infty\}$ be convex and $z \in \operatorname{dom} f$ then*

$$f(z) = \min_{x \in X} f(x) \quad \Leftrightarrow \quad 0 \le f(x) - f(z) \quad \forall x \in X \quad \Leftrightarrow \quad 0 \in \partial f(z).$$

Therefore, $0 \in \partial f(z)$ is the generalization of the classical optimality condition $0 = f'(z)$. The subgradient is also the generalization of the gradient in the sense that if f is Gâteaux-differentiable, then $\partial f(x) = \{\nabla f(x)\}$.

In general the subgradient of a function may also be the empty set. However, in this thesis we will deal with norms on Banach spaces which are Lipschitz-continuous with respect to itself. Hence, the subgradient of a Banach space norm is not empty, cf. [62, II.7.5].

To derive the optimality conditions for the Tikhonov functional (2.1), which is a sum of two powers of norms, we have to know what the subgradient of a sum of two convex functions is. Let f and g be two convex functions, then

$$\partial(f + g)(x) \supset \partial f(x) + \partial g(x)$$

since for any $x^* \in \partial f(x)$ and $z^* \in \partial g(x)$ we have

$$f(x) + \langle x^*, y - x \rangle + g(x) + \langle z^*, y - x \rangle \leq f(y) + g(y).$$

The next theorem provides some sufficient conditions on f and g to satisfy $\partial(f + g)(x) \subset \partial f(x) + \partial g(x)$ too.

Theorem 2.19 (Subgradient of sums). *[73, Theorem 47.B] Let X be a Banach space and $f, g : X \to \mathbb{R} \cup \{\infty\}$ be convex. If there is a point in $(\mathrm{dom}\, f) \cap (\mathrm{dom}\, g)$, where f is continuous, then for all $x \in X$ it holds*

$$\partial(f + g)(x) = \partial f(x) + \partial g(x).$$

Theorem 2.20 (Subgradient of compositions). *[62, II.7.8] Let $f : Y \to \mathbb{R} \cup \{\infty\}$ be convex and $A : X \to Y$ be continuous and linear, and assume f is continuous at some point of the range of A. Then*

$$\partial(f \circ A)(x) = A^*(\partial f(Ax)).$$

Theorem 2.21 (Subgradient of translation). *Let $f : X \to \mathbb{R} \cup \{\infty\}$ be convex. Then*

$$\partial(f(\cdot - y))(x) = (\partial f)(x - y).$$

Proof. Let $x^* \in \partial(f(\cdot - y))(x)$, then

$$f(z - y) \geq f(x - y) + \langle x^*, z - x) \rangle, \qquad \forall z \in X,$$

hence with $a = z - y$

$$f(a) \geq f(x - y) + \langle x^*, a - (x - y) \rangle \qquad \forall a \in X.$$

Therefore, we get $\partial(f(\cdot - y))(x) \subset (\partial f)(x - y)$. In the same way one can prove $\partial(f(\cdot - y))(x) \supset (\partial f)(x - y)$. □

As a consequence of the last three theorems, we are able to write down the optimality conditions for the Tikhonov functional (2.1). The next corollary is therefore the main result of this section.

Corollary 2.22. *We have*

$$\partial(\tfrac{1}{p}\|A \cdot -y^\delta\|_Y^p)(x) = (A^* \partial \tfrac{1}{p}\| \cdot -y^\delta\|_Y^p)(Ax) = (A^* \partial \tfrac{1}{p}\| \cdot \|_Y^p)(Ax - y^\delta).$$

Hence, if x_α^δ minimizes the functional T_α, then

$$0 \in A^*(\partial \tfrac{1}{p}\| \cdot \|_Y^p)(Ax_\alpha^\delta - y^\delta) + \alpha(\partial \tfrac{1}{q}\| \cdot \|_X^q)(x_\alpha^\delta). \tag{2.2}$$

As last fact from general convex analysis we mention that the subgradient is also the right notion to generalize the concept of monotonicity. We recall that if $f : \mathbb{R} \to \mathbb{R}$ is differentiable and convex then the derivative f' is monotone, i.e.

$$(f'(x) - f'(y)) \cdot (x - y) \geq 0 \qquad \forall x, y.$$

The next theorem shows that this fact is also true for all convex mappings if the derivative is replaced by the subgradient.

Theorem 2.23. *[54, Theorem 12.17] The subgradient of a proper, convex mapping f is monotone, i.e.*

$$\langle x^* - y^*, x - y \rangle \geq 0 \qquad \forall x^* \in \partial f(x), y^* \in \partial f(y).$$

2.2.1 Duality mapping

In Corollary 2.22 we derived the optimality condition for the Tikhonov functional (2.1). This condition employs the mappings

$$\partial(\tfrac{1}{p} \| \cdot \|_Y^p) \qquad \text{and} \qquad \partial(\tfrac{1}{q} \| \cdot \|_X^q).$$

To effectively make use of the optimality condition we need more information about the structure of the above subgradients of powers of Banach space norms. We first introduce the so-called duality mapping J_p^X.

Definition 2.24 (Duality mapping). *The (set-valued) mapping $J_p^X : X \rightrightarrows X^*$ with $p > 1$ defined by*

$$J_p^X(x) := \{x^* \in X^* \ : \ \langle x^*, x \rangle = \|x\| \|x^*\|, \|x^*\| = \|x\|^{p-1}\}$$

is called the duality mapping of X with gauge function $t \mapsto t^{p-1}$.

We will denote by j_p^X a single valued selection of J_p^X, i.e. $j_p^X : X \to X^*$ is a mapping with $j_p^X(x) \in J_p^X(x)$ for all $x \in X$. We will sometimes abbreviate this to $j_p^X \in J_p^X$.

However, for single valued duality mappings J_p^X we will use the symbols J_p^X and j_p^X interchangeably, i.e. formally $j_p^X \equiv J_p^X$.

Finally, we stress that the duality mapping should not be confused with the dual mapping introduced in Definition 2.3. As usual we will omit the index indicating the space, where it is clear from the context which one is meant.

The good news is: the duality mapping is the desired subgradient of powers of Banach norms.

Theorem 2.25 (Theorem of Asplund). *[3] Let X be a normed space and $p > 1$ then*

$$J_p^X = \partial \left(\tfrac{1}{p} \| \cdot \|_X^p \right).$$

With the definition of the duality mapping J_p^X and its characterization by the Theorem of Asplund (Theorem 2.25) we have already some tools at hand to describe the subgradients of powers of norms. However, what we still lack is the exact form of J_p^X. In the next example we provide J_p^X in an explicit form for the case of Hilbert spaces and sequence spaces.

Example 2.26. *In Hilbert spaces we have $J_2(x) = x$ and therefore*

$$J_p(x) = \|x\|^{p-2} \cdot x.$$

Next, we consider the usual sequence norms on ℓ^r spaces defined via

$$\|x\|_r := \left(\sum_i |x_i|^r \right)^{1/r}$$

equipped with the duality pairing $\langle x, y \rangle = \sum_i x_i y_i$. Here, we use r for the exponent of the sequence spaces, since in Banach space theory the letter p is usually used for the exponent of the duality mapping, resp. the power of the norm.
   Then, for $1 < r < \infty$ we can easily compute

$$(J_p(x))_i := ((\nabla \tfrac{1}{p} \| \cdot \|_r^p)(x))_i = \|x\|_r^{p-r} |x_i|^{r-1} \operatorname{sign}(x_i).$$

We can write ∇ instead of ∂ since the functional is smooth, cf. [59, 45, 71].
   For $r = 1$ we have

$$(J_p(x))_i := (\partial \tfrac{1}{p} \| \cdot \|_1^p)_i = \|x\|_1^{p-1} \cdot \operatorname{Sign}(x_i)$$

where Sign is the set-valued sign function, defined in the same way as usual sign function except that $\operatorname{Sign}(0) = [-1, 1]$, cf. [46, Corollary 3.3].
   We do not consider the space $r = \infty$ since it is already difficult to identify the dual space of ℓ^∞ and the subgradient of its norm in a closed way. The situation is much simpler if we consider only the finite dimensional cases, i.e. $\mathbb{R}^N$ equipped with the $\| \cdot \|_\infty$ norm. Then, the dual space is the finite dimensional version of ℓ^1 and

$$(J_p(x))_i = \|x\|_\infty^{p-1} \cdot y_i$$

where $y_i = 0$ for i such that $|x_i| \neq \|x\|_\infty$, $\operatorname{sign}(y_i) = \operatorname{sign}(x_i)$ for all other i and

$$\sum_{i=1}^{N} |y_i| = 1.$$

One of the elements of the subgradient is given by $y_i = sign(x_i)$ for exactly one i such that $|x_i| = \|x\|_\infty$ and $y_i = 0$ for all other i. This is the choice employed in e.g. [59, 61].

Remark 2.27. *The above example shows that - at least for the sequence spaces - the numeric complexity (i.e. the number of float operations) of the evaluation of the norm is comparable to the evaluation of the duality mapping and is about of order $\mathcal{O}(N)$, where N is the dimension of the space. Further, the evaluation of the duality mapping is not slower than the evaluation of an usual operator mapping from $\mathbb{R}^N$ to $\mathbb{R}^N$, which we assume to be at least of order $\mathcal{O}(N)$ resp. $\mathcal{O}(N \cdot \log N)$ float operations for the so-called fast operators and $\mathcal{O}(N^2)$ for all other operators.*

In all the algorithms presented in this work we will always make the quiet assumption that the evaluation of the norm and the duality mapping is computationally not more expensive than the evaluation of the operator.

To present the most important properties of the duality mapping we need some notions from the geometry of Banach spaces. Therefore, we will first introduce the necessary notions in the next section and then state the properties of the duality mapping in Theorem 2.44.

2.3   Geometry of Banach space norms

Is there any way to say that a Banach space is "nice" from a geometrical point of view? We know that the norm of a Banach space is convex, hence also the powers of norms are convex (assuming the exponent is not smaller than one). So far, the only geometrical notion connected to convex functions is the subgradient. Therefore, it is natural to introduce "niceness" of a Banach space by improving (or enhancing) the subgradient definition. Further, we keep in mind that the subgradient of a power of the norm is the duality mapping.

Altogether we know that for all $z \in X$ we have

$$\tfrac{1}{p}\|z\|^p - \tfrac{1}{p}\|x\|^p - \langle J_p(x), z - x \rangle \geq 0,$$

resp. with $y = -(z - x)$ we have

$$\tfrac{1}{p}\|x - y\|^p - \tfrac{1}{p}\|x\|^p + \langle J_p(x), y \rangle \geq 0.$$

We are interested in upper and lower bounds of the left-hand side of the above inequality in terms of norm of y, say

$$\tfrac{1}{p}\|x - y\|^p - \tfrac{1}{p}\|x\|^p + \langle J_p(x), y \rangle \geq \tfrac{c_p}{p}\|y\|^p$$

or

$$\tfrac{G_p}{p}\|y\|^p \geq \tfrac{1}{p}\|y - x\|^p - \tfrac{1}{p}\|x\|^p + \langle J_p(x), y \rangle.$$

for some $p, G_p, c_p > 0$.

We will continue as follows: First, in Definitions 2.28 and 2.29, we postulate the above inequalities as properties of Banach spaces. We will call the postulated properties convexity of power type and smoothness of power type. Next, in Examples 2.37 and 2.38, we will see that most of the common spaces are convex and/or smooth of power type.

We will also see that if a space is convex of power type then the dual space is smooth of power type. And if a space is smooth of power type then the dual space is convex of power type, cf. Theorem 2.43, which shows that the notions of convexity and smoothness are - in a certain sense - dual.

With all the necessary notions from the geometry of Banach spaces at hand we will be ready to state the most important properties of the duality mapping.

Finally, we will introduce the so-called Bregman distance and show how the Bregman distance relates to the norm (Definition 2.46 and Theorem 2.48).

We notice that the duality between convexity of power type and smoothness of power type (cf. Theorem 2.43), the transportation property of the duality mappings (Theorem 2.44) and the relation between the Bregman distance and the norm (Theorem 2.48) are the three essential ingredients in the convergence analysis of the algorithms introduced in later sections.

2.3.1   Convexity and smoothness of power type

We start with definitions of p-convexity and p-smoothness:

Definition 2.28. *We call X convex of power type p or p-convex if there exists a constant $c_p > 0$ such that*

$$\tfrac{1}{p}\|x - y\|^p \geq \tfrac{1}{p}\|x\|^p - \langle j_p(x), y \rangle + \tfrac{c_p}{p}\|y\|^p$$

for all $x, y \in X$ and all $j_p \in J_p$.

Definition 2.29. *We call X smooth of power type p or p-smooth if there exists a constant $G_p > 0$ such that*

$$\tfrac{1}{p}\|x - y\|^p \leq \tfrac{1}{p}\|x\|^p - \langle j_p(x), y \rangle + \tfrac{G_p}{p}\|y\|^p$$

for all $x, y \in X$ and all $j_p \in J_p$.

Remark 2.30. *We notice that the most interesting case in the above definition is $y \to 0$ since the subgradient can be considered as a generalization of the derivative and in the definition of derivative the main case is $y \to 0$.*

1. *Let X be p-convex. The smaller p is, the bigger the term $\|y\|^p$ gets for $y \to 0$. Geometrically this means that the functional is not flat, which is a good thing. Our understanding is therefore: for p-convexity the smaller p is the better.*

2. *The opposite is true for p-smoothness. Let X be p-smooth. Then, the estimate in the definition is the better, the bigger p is, since the bigger p is, the faster $\|y\|^p$ vanishes for $y \to 0$. Hence, our understanding for p-smoothness is: the bigger p is the better.*

Remark 2.31. *Notice that if a space is p-smooth for some $p > 1$, then the p-th power of norm is already Fréchet differentiable, hence Gâteaux differentiable and therefore $J_p(x)$ is single valued for all x. By Theorem 2.44 we see that if $J_p(x)$ is single-valued for all x and some $p > 1$ then $J_p(x)$ is single-valued for all x and all $p > 1$.*

Xu and Roach [71] have proven the following characterizations of convexity and smoothness of power type, which will be very useful in some proofs of our algorithms.

Theorem 2.32 (Xu-Roach inequalities). *The following statements are all equivalent:*

1. X *is δ-convex.*

2. *For some p such that $1 < p < \infty$, some $j_p \in J_p$ and for every $x, y \in X$ we have*
$$\langle j_p(x) - j_p(y), x - y \rangle \geq C \max\{\|x\|, \|y\|\}^{p-\delta} \|x - y\|^\delta.$$

3. *The last statement holds for all p such that $1 < p < \infty$ and all $j_p \in J_p$.*

4. *For some p such that $1 < p < \infty$, some $j_p \in J_p$ and for every $x, y \in X$ we have*
$$\|x - y\|^p \geq \|x\|^p - p\langle j_p(x), y \rangle + \sigma_p(x, y)$$
where
$$\sigma_p(x, y) \geq C \int_0^1 t^{\delta-1} \max\{\|x - ty\|, \|x\|\}^{p-\delta} \|y\|^\delta dt.$$

5. *The last statement holds for all p such that $1 < p < \infty$ and all $j_p \in J_p$.*

The generic constant $C > 0$ can be chosen independently of x and y. Further, if X is δ-convex with constant c_δ in the definition of convexity of power type, then for every $p > 1$ there exist numbers $k_{p,\delta,2} > 0$ and $k_{p,\delta,4} > 0$, such that the constants in the second and fourth item may be chosen as $k_{p,\delta,2} \cdot c_\delta$ resp. $k_{p,\delta,4} \cdot c_\delta$.

Theorem 2.33 (Xu-Roach inequalities). *The following statements are all equivalent:*

1. X *is ρ-smooth.*

2. *For some p such that $1 < p < \infty$ J_p is single-valued and for all $x, y \in X$ we have*
$$\|J_p(x) - J_p(y)\| \leq C \max\{\|x\|, \|y\|\}^{p-\rho} \|x - y\|^{\rho-1}.$$

3. *The last statement holds for all all p such that $1 < p < \infty$.*

4. *For some some p such that $1 < p < \infty$, some $j_p \in J_p$ and for all $x, y \in X$ we have*
$$\|x - y\|^p \leq \|x\|^p - p\langle j_p(x), y \rangle + \sigma_p(x, y),$$
where
$$\sigma_p(x, y) \leq C \int_0^1 t^{\rho-1} \max\{\|x - ty\|, \|x\|\}^{p-\rho} \|y\|^\rho dt.$$

5. *The last statement holds for all p such that $1 < p < \infty$ and all $j_p \in J_p$.*

The generic constant $C > 0$ can be chosen independently of x and y. Further, if X is ρ-smooth with constant G_ρ in the definition of smoothness of power type, then for every $p > 1$ there exist numbers $K_{p,\rho,2} < \infty$ and $K_{p,\rho,4} < \infty$, such that the constants in the second and third item may be chosen as $K_{p,\delta,2} \cdot G_\rho$ resp. $K_{p,\delta,4} \cdot G_\rho$.

Corollary 2.34. *Let X be p-smooth then for all q such that $1 < q < p$ the space X is also q-smooth. Of course a similar result holds for the convexity of power type: If X is p-convex then for all $p < q < \infty$ the space X is also q-convex.*

Proof. By the Xu-Roach characterization of smoothness of power type in Theorem 2.33 we have for p-smooth space and $q > 1$ that

$$\|J_q(x) - J_q(y)\| \le C \max\{\|x\|, \|y\|\}^{q-p} \|x - y\|^{p-1}.$$

Due to $q < p$ we get

$$\max\{\|x\|, \|y\|\}^{q-p} \|x - y\|^{p-1} \le C \|x - y\|^{q-1}$$

and therefore

$$\|J_q(x) - J_q(y)\| \le C \|x - y\|^{q-1}.$$

Hence, again by the Xu-Roach characterization of smoothness of power type the space X is q-smooth. The proof for convexity of power type is analogue. $\qquad\square$

Corollary 2.35. *Let X be ρ-smooth and $p > 1$. Then, the duality mapping J_p is $\min\{p-1, \rho-1\}$-Hölder continuous on bounded sets and single valued.*

Proof. If $p \le \rho$ then by Corollary 2.34 we know that X is p-smooth too. Therefore, by Theorem 2.33 the duality mapping J_p is $(p-1)$ Hölder continuous. If $\rho \le p$ then by Theorem 2.33 we have

$$\|J_p(x) - J_p(y)\| \le C \max\{\|x\|, \|y\|\}^{p-\rho} \|x - y\|^{\rho-1} \le C \|x - y\|^{\rho-1}$$

on bounded sets, which proves the claim. $\qquad\square$

Remark 2.36. *The definition of convexity of power type in Definition 2.28 and smoothness of power type in Definition 2.29 can be regarded as a special case of the Theorems 2.32 and 2.33 where the index p of the duality mapping has the same value as the index δ resp. ρ denoting the type of power of the space.*

Next, we present prominent examples of spaces convex of power type and smooth of power type.

Example 2.37. *In Hilbert spaces the polarization identity*

$$\tfrac{1}{2}\|x - y\|^2 = \tfrac{1}{2}\|x\|^2 - \langle x, y \rangle + \tfrac{1}{2}\|y\|^2$$

ensures that Hilbert spaces are 2-convex and 2-smooth.

Example 2.38. *Let* $\Omega \subset \mathbb{R}^n$ *be a domain. It is known [71, 45] that*

- *the sequence spaces* ℓ^r

- *the Lebesgue spaces* $L^r(\Omega)$

- *the Sobolev spaces* $W^r_m(\Omega)$

with the usual norms and

$$1 < r < \infty$$

are

$$\max\{2, r\} - convex$$

and

$$\min\{2, r\} - smooth.$$

For ℓ^1 *one can show with* $x = (1, 0, \ldots), j_p(x) = (1, 0, \ldots), y = (0, 0, \ldots), j_p(y) = (1, 0, \ldots)$ *that the space cannot be p-convex or p-smooth for any p.*

The examples above justify our view that convexity and smoothness of power type is quite common. Further, we see that the powers in the convexity and smoothness of power type interpolate between the space ℓ^2, which as a Hilbert space may be regarded as a geometrically well-behaved space, and the spaces like ℓ^1 or ℓ^∞, which are regarded as geometrically not so nice, since no direct generalizations of the polarization identity can be established in these spaces.

In Example 2.37 and 2.38 we saw that most of the usual spaces are smooth and convex of some power type. But this is not the end of the good news. In fact one can show that even spaces uniformly convex *or* uniformly smooth are (up to some equivalent renorming) also smooth *and* convex of power type. We have:

Theorem 2.39. *[16, Theorem 5.2] Let* X *be uniformly convex, i.e. for the* modulus of convexity $\delta_X : [0, 2] \to [0, 1]$ *defined via*

$$\delta_X(\epsilon) := \inf\left\{1 - \left\|\frac{1}{2}(x + y)\right\| : \|x\| = \|y\| = 1, \|x - y\| \geq \epsilon.\right\}$$

we have

$$\delta_X(\epsilon) > 0 \qquad \forall 0 < \epsilon \leq 2$$

or let X *be uniformly smooth, i.e. for the modulus of smoothness* $\rho_X : [0, \infty) \to [0, \infty)$ *defined via*

$$\rho_X(\tau) := \frac{1}{2}\sup\{\|x + y\| + \|x - y\| - 2 : \|x\| = 1, \|y\| \leq \tau\}$$

we have

$$\lim_{\tau \to 0} \frac{\rho_X(\tau)}{\tau} = 0,$$

then there exists an equivalent norm, such that X equipped with this norm is smooth of power type and convex of power type.

Further, every uniformly convex and every uniformly smooth space is reflexive.

Remark 2.40. *We remark that although the proof of this theorem is constructive, it relies heavily on Banach space geometry, especially notions of superreflexivity and Rademacher-type.*

In the the light of the above facts we consider in what follows only spaces smooth and convex of power type. These spaces share many very interesting properties, which we summarize in the following theorems:

Theorem 2.41. *Let X be p-convex then (cf. [71, p. 193], [11, Chapter II])*

1. $p \geq 2$,

2. X *is uniformly convex and the modulus of convexity is given by* $\delta_X(\epsilon) \geq C\epsilon^p$,

3. X *is strictly convex, i.e* $\left\| \frac{1}{2}(x+y) \right\| < 1$ *for all* $x \neq y$ *on the unit ball,*

4. X *is reflexive.*

Theorem 2.42. *Let X be p-smooth then (cf. [71, p. 193], [11, Chapter II])*

1. $p \leq 2$,

2. X *is uniformly smooth and the modulus of smoothness of X is given by* $\rho_X(\tau) \leq C\tau^p$,

3. X *is smooth, i.e.* $J_p^X(x)$ *is single-valued for all* $x \in X$,

4. X *is reflexive.*

The next theorem allows us to connect properties of the primal space with the properties of the dual space.

Theorem 2.43 (Duality of convexity and smoothness). *[45, Vol II 1.e] We have:*

1. X *is p-smooth if and only if X^* is p^*-convex.*

2. X *is p-convex if and only if X^* is p^*-smooth.*

Finally, we are ready to present the most important properties of the duality mapping:

Theorem 2.44. *We have (cf. [59, Lemma 2.3 and 2.5], [11, Proposition I.4.7.f and II.3.6] and [71]):*

1. *For every $x \in X$ the set $J_p(x)$ is non-empty and convex.*

2. $J_p(-x) = -J_p(x)$ and $J_p(\lambda x) = \lambda^{p-1} J_p(x)$ for all $x \in X$ and all $\lambda > 0$.

3. For $p, q > 1$ we have

$$\|x\|^{q-1} J_p(x) = \|x\|^{p-1} J_q(x).$$

4. If X is convex of power type and smooth then J_p^X is single valued, norm-to-weak continuous, bijective and the duality mapping $J_{p^*}^{X^*}$ is single valued and

$$J_{p^*}^{X^*}(J_p^X(x)) = x.$$

Remark 2.45. The first claim is a consequence of the Hahn-Banach theorem, the next two claims are straightforward applications of the definition of the duality mapping. The last claim holds under the weaker condition that X is smooth, strictly convex and reflexive.

The importance of the last fact can be hardly overestimated, since it states that for nice enough spaces (e.g. being smooth of power type and convex of power type) the duality mappings on the primal space and the dual space can be used to transport all elements from the primal to dual space and vice versa. We have

$$J_{p^*}^{X^*} J_p^X(x) = x \qquad \text{and} \qquad J_p^X J_{p^*}^{X^*}(x^*) = x^* \qquad \forall x \in X \quad \text{and} \quad \forall x^* \in X^*.$$

2.3.2 Bregman distances

Due to geometrical properties of Banach spaces it is often more appropriate to use the so-called Bregman distance instead of functionals like $\|x - y\|^p$ or $\|j_p(x) - j_p(y)\|^p$ to prove convergence of algorithms. The main idea of the Bregman distance is to use the gap between a function and its linearization instead of the function itself to measure distances.

Definition 2.46 (Bregman distance). Let $j_p : X \to X^*$ be a single valued selection of the duality mapping J_p. Then, the functional

$$D_{j_p}(x, y) = \tfrac{1}{p}\|y\|^p - \tfrac{1}{p}\|x\|^p - \langle j_p(x), y - x \rangle$$

is called the Bregman distance (of $\frac{1}{p}\| \cdot \|^p$).

Remark 2.47. First, we remark that - to the best of author's knowledge - the notion of Bregman distances was introduced by Bregman in [9].

Further, we remark that there are several possible ways to define the Bregman distance of a convex functional. However, the main idea is always to measure the gap between the functional and its linearization. We visualize this concept in Figure 2.2.

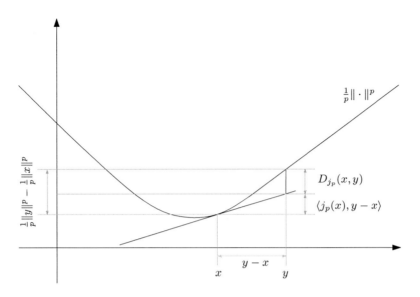

Figure 2.2: Geometrical interpretation of the Bregman distance as the gap between the function and its linearization.

The definition of the Bregman distance in this work is the same as in [61, 59]. However, we must warn the reader that there are authors - e.g. [35, 10] - which define the Bregman distance via

$$d_{j_p}(x, y) = \tfrac{1}{p}\|x\|^p - \tfrac{1}{p}\|y\|^p - \langle j_p(y), x - y \rangle$$

which is slightly different to the Definition 2.46. Namely the arguments are interchanged, i.e. the second argument in [35, 10] is the first in this thesis and vice versa.

Finally, we remark that the Bregman distance at some point is only defined if the subgradient at this point is not empty. However, the subgradient of power of Banach space norms is never empty, cf. Theorem 2.44.

Theorem 2.48. Let X be a Banach space, $j_p \in J_p$ a fixed single valued selection of the duality mapping J_p, then (cf. [59, Theorem 2.12] and [71]):

1. $D_{j_p}(x, y) \geq 0$.

2. $D_{j_p}(x, y) = 0$ if and only if $j_p(x) \in J_p(y)$.

3. If X is convex of power type, then $(x_n)_n$ remains bounded if $(D_{j_p}(x_n, y))_n$ is bounded.

4. $D_{j_p}(x, y)$ is continuous in its second argument. If X is convex of power type, then J_p is continuous on bounded subsets and $D_{j_p}(x, y)$ is continuous in its first argument.

5. If X is convex of power type then the following are equivalent:

   (a) $\lim_{n\to\infty} \|x_n - x\| = 0$,
   (b) $\lim_{n\to\infty} \|x_n\| = \|x\|$ and $\lim_{n\to\infty} \langle J_p(x_n), x \rangle = \langle J_p(x), x \rangle$,
   (c) $\lim_{n\to\infty} D_{j_p}(x_n, x) = 0$.

6. $(x_n)_n$ is a Cauchy sequence if it is bounded and for all $\epsilon > 0$ there is an $n_0 \in \mathbb{N}$ such that $D_{j_p}(x_k, x_l) < \epsilon$ for all $k, l \geq n_0$.

7. X is p-convex if and only if $D_{j_p}(x, y) \geq C\|x - y\|^p$,

8. X is p-smooth if and only if $D_{j_p}(x, y) \leq C\|x - y\|^p$.

Next, we formulate corollaries of the Xu-Roach inequalities, cf. Theorem 2.32, regarding the Bregman distances:

Corollary 2.49. *Let X be δ-convex, then with the function σ_p of the Xu-Roach characterization of the convexity of power type in Theorem 2.32 we have:*

1. *If $1 < p \leq \delta$ then*

$$D_{j_p}(x, y) \geq C \cdot \sigma_p(x, x - y) \geq C \cdot (\|x\| + \|y\|)^{p-\delta} \|x - y\|^\delta.$$

2. *If $\delta \leq p < \infty$ then*

$$D_{j_p}(x, y) \geq C \cdot \sigma_p(x, x - y) \geq C \cdot \|x - y\|^p.$$

The generic constant $C > 0$ can be always chosen independently of x and y.

Proof. By the Xu-Roach characterization of convexity of power type in Theorem 2.32 we have

$$
\begin{aligned}
D_{j_p}(x, y) &= \tfrac{1}{p}\|y\|^p - \tfrac{1}{p}\|x\| - \langle j_p(x), y - x \rangle \\
&\geq \sigma_p(x, x - y) \\
&\geq C \int_0^1 t^{\delta-1}(\max\{\|x - t(x - y)\|, \|x\|\}^{p-\delta} \|x - y\|^\delta \, dt
\end{aligned}
$$

then the claims are a consequence of the fact that

$$\|x\| + \|y\| \geq \max\{\|x - t(x - y)\|, \|x\|\} \geq \tfrac{t}{2}\|x - y\|$$

for $0 \leq t \leq 1$. $\qquad\square$

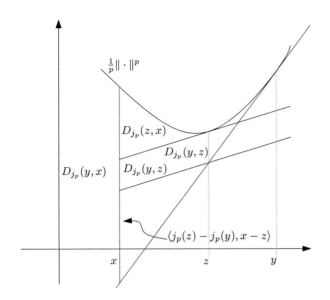

Figure 2.3: Geometrical interpretation of the three-point identity, cf. Lemma 2.3.

We again return to the properties of the Bregman distances. The following identity for Bregman distances is known as the *three-point identity:*

Lemma 2.50. *Let j_p be a single valued selection of the duality mapping J_p^X then we have*

$$D_{j_p}(y, x) = D_{j_p}(z, x) + D_{j_p}(y, z) + \langle j_p(z) - j_p(y), x - z \rangle.$$

Proof. We have

$$
\begin{aligned}
D_{k_p}(z, x) \quad + \quad & D_{j_p}(y, z) + \langle j_p(z) - j_p(y), x - z \rangle \\
= \quad & \tfrac{1}{p}\|x\|^p - \tfrac{1}{p}\|z\|^p - \langle j_p(z), x - z \rangle \\
& + \tfrac{1}{p}\|z\|^p - \tfrac{1}{p}\|y\|^p - \langle j_p(y), z - y \rangle \\
& + \langle j_p(z), x - z \rangle - \langle j_p(y), x - z \rangle \\
= \quad & \tfrac{1}{p}\|x\|^p - \tfrac{1}{p}\|y\|^p - \langle j_p(y), x - y \rangle \\
= \quad & D_{j_p}(y, x).
\end{aligned}
$$

□

We notice that the three-point identity has a geometric interpretation, as shown in Figure 2.3.

We also remark that there is a close connection between the primal Bregman distances $D_{j_p}^X$ and the related dual Bregman distances $D_{j_{p^*}}^{X^*}$.

Lemma 2.51. *Let j_p^X be a single valued selection of the duality mapping J_p^X. If there exists a single valued selection $j_{p^*}^{X^*}$ of $J_{p^*}^{X^*}$ such that for some fixed $y \in X$ we have $j_{p^*}^{X^*}(j_p^X(y)) = y$ then*

$$D_{j_p}^X(x, y) = D_{j_{p^*}}^{X^*}(j_p^X(y), j_p^X(x))$$

for all $x \in X$.

Proof. We omit the indices indicating the spaces. By a straightforward computation we get

$$
\begin{aligned}
D_{j_p}(x, y) &= \tfrac{1}{p}\|y\|^p - \tfrac{1}{p}\|x\|^p - \langle j_p(x), y - x \rangle \\
&= \tfrac{1}{p^*}\|x\|^p - \tfrac{1}{p^*}\|y\|^p - \langle y, j_p(x) - j_p(y) \rangle.
\end{aligned}
$$

For all $x \in X$ we have $\|x\|^p = \|x\|^{(p-1)\cdot p^*} = \|j_p(x)\|^{p^*}$. Therefore, we have

$$D_{j_p}(x, y) = \tfrac{1}{p^*}\|j_p(x)\|^{p^*} - \tfrac{1}{p^*}\|j_p(y)\|^{p^*} - \langle y, j_p(x) - j_p(y) \rangle.$$

The last right-hand side is equal to $D_{j_{p^*}}(j_p(y), j_p(x))$, since by assumption $y = j_{p^*}(j_p(y))$. □

As a consequence of the above lemmas we get the following form of the three-point identity:

Theorem 2.52. *Let X be convex of power type and smooth of power type, further let $y = j_{p^*}(y^*)$ and $z = j_{p^*}(z^*)$, where j_{p^*} is the unique single-valued selection of $J_{p^*}^{X^*}$. Then, for the unique single-valued selection j_p of J_p^X we have*

$$D_{j_p}(y, x) = D_{j_p}(z, x) + D_{j_{p^*}}(z^*, y^*) + \langle z^* - y^*, x - z \rangle.$$

Proof. Since X is convex of power type and smooth of power type the dual X^* is, by Theorem 2.43, convex of power type and smooth of power type too. Therefore, by Theorem 2.33 the mappings J_p^X and $J_{p^*}^{X^*}$ have only one single valued selection, say j_p and j_{p^*} and by Theorem 2.44 $j_p(j_{p^*}(x^*)) = x^*$ for all $x^* \in X^*$ and $j_{p^*}(j_p(x)) = x$ for all $x \in X$. Therefore, $j_p(y) = j_p(j_{p^*}(y^*)) = y^*$ and $j_p(z) = j_p(j_{p^*}(z^*)) = z^*$ and due to Lemma 2.51 we have $D_{j_p}(y, z) = D_{j_{p^*}}(j_p(z), j_p(y)) = D_{j_{p^*}}(z^*, y^*)$. The claim is then a consequence of the three-point identity of Lemma 2.3. □

2.4   Regularization theory

Let X and Y be two Banach spaces and $A : X \to Y$ a linear, continuous operator and $y \in \mathcal{R}(A)$. If A is not injective then the problem

$$Ax = y$$

has more than one solution. In such case one is usually interested in the solution which has the smallest norm. Therefore, we define:

Definition 2.53 (Minimum norm solution). *Let X and Y be two Banach spaces, $A : X \to Y$ a linear, continuous operator and $y \in \mathcal{R}(A)$. We call $x^\dagger \in X$ minimum norm solution of the problem $Ax = y$ if*

$$Ax^\dagger = y \qquad \text{and} \qquad \|x^\dagger\| = \inf\{\|z\| \ : \ Az = y\}.$$

Remark 2.54. *Notice that in the setting of Hilbert spaces, i.e. X and Y being Hilbert spaces the term minimum norm solution is sometimes used to describe the solution with minimal norm of the normal equations $A^*Ax = A^*y$. However, this is not the sense in which we use this term here.*

The following lemma gives us an important characterization of the minimum norm solution:

Lemma 2.55. *[59, Lemma 2.10] Let X be a Banach space convex of power type, Y be an arbitrary Banach space, $A : X \to Y$ be linear and continuous, $y \in \mathcal{R}(A)$ then the minimal norm solution $x^\dagger$ of $Ax = y$ exists and is unique. If additionally the space X is smooth of power type and for some $x \in X$ and $q > 1$ we have $j_q(x) \in \overline{\mathcal{R}(A^*)}$ and $x - x^\dagger \in \mathcal{N}(A)$ then $x = x^\dagger$.*

Definition 2.56 (Regularization). *The mapping*

$$Y \times \mathbb{R} \to X \qquad \text{with} \qquad (y^\delta, \alpha) \mapsto x_\alpha^\delta$$

is called a regularization of the linear operator equation $Ax = y$ with $y \in \mathcal{R}(A)$ if for every sequence (y_k) with $\|y_k - y\| \leq \delta_k$ and $\delta_k \to 0$ as $k \to \infty$ there exists a sequence (α_k) such that $x_{\alpha_k}^{\delta_k}$ converge to the minimum norm solution $x^\dagger$ of the problem $Ax = y$.

By Bakushinskii's Veto [4] we know that the parameter α_k cannot be chosen independently on the noise level δ_k. We call a parameter choice a-priori if the parameter α_k depends only on δ_k. Further, we call a parameter choice a-posteriori if the parameter α_k depends on δ_k as well as y_k.

3

Geometry of Besov spaces

The main aim of this chapter is to prove that the characterization of Besov norms via wavelet coefficients is convex of power type and smooth of power type.

Since their introduction in the 1980s wavelets have conquered the field of signal processing. Especially the discovery of smooth wavelets with compact support by Daubechies [14] and the development of fast algorithms in the 1990s have given this decomposition method an tremendous boost. Today, aside of the Fourier transform the wavelet decomposition is probably the most used tool for analyzing smoothness of functions or signals.

In a formal way, if we denote by $\mathcal{F}f$ the Fourier transform of f, cf. (3.1), then

$$\mathcal{F}(\mathrm{D}\,f)(\omega) = (i\omega)\mathcal{F}(f)(\omega),$$

where D is the differentiation operator. Hence, it is possible to describe the smoothness of a function by the properties of its Fourier transform. The norms arising in such a spectral description of smoothness are the so-called fractional Sobolev norms, which we will discuss in the next section. However, the trigonometric functions used in the computation of the Fourier transform are not localized and therefore little (if any) information about local properties (e.g. support) of a function can be retrieved from its Fourier transform.

The main advantage of the wavelet decomposition over the Fourier transform is: wavelets are usually constructed to have compact support and therefore are by construction well-localized. As a consequence local properties of a function can easily be retrieved from its wavelet coefficients.

So, in the analogy to the Fourier case one could ask, whether the smoothness of a function can be measured via it's wavelet coefficients. The answer is yes and the spaces induced by the (discrete) wavelet transform are the so-called Besov spaces, which we will introduce in Section 3.2. More information on use of Besov spaces in applications like image processing or inverse problems can be found in [46] and [15] and the references therein. We also remark that Besov spaces are also closely related to N-best term approximation with means of wavelets, cf. [17].

Besov spaces can be introduced in many equivalent ways in the sense that all of these approaches generate the same topology. However, not every approach is equivalent in the sense of geometrical properties of the norm. The approach via wavelet-expansions is probably the most appealing one, since at the same time fast routines for computing wavelet coefficients are available today and therefore evaluating the norm is not computationally challenging *and* at the same time the associated norms have good geometrical properties.

3.1 Sobolev spaces

Sobolev spaces are widely known and one of the most used function spaces. They were introduced by Sobolev in the mid-thirties, see [63, 64, 65]. Sobolev spaces are a very powerful tool in functional analysis and its applications in partial differential equations. Therefore, today they are widely accepted and used, cf. e.g. [1, 49, 50]. For the sake of simplicity and clarity we only consider Sobolev spaces on the one-dimensional real line, since the extension to the multi-dimensional case is straight-forward, c.f. [66, 67, 68].

Let $k \in \mathbb{N}$ and $1 < p < \infty$ then the Sobolev spaces $W_p^k(\mathbb{R})$ are defined by

$$W_p^k(\mathbb{R}) = \{f \in L^p(\mathbb{R}) \ : \ \mathrm{D}^\alpha f \in L^p(\mathbb{R}), \forall \alpha \leq k\}$$

where D^k is the k-th distributional derivative (for exact definition cf. [1]). The norms on these spaces may be defined via

$$\|f\|_{W_p^k(\mathbb{R})} = \|f\|_{L^p(\mathbb{R})} + \sum_{\alpha \leq k} \|\mathrm{D}^\alpha f\|_{L^p(\mathbb{R})}.$$

We note that the spaces $W_p^k(\mathbb{R})$ form a scale, i.e.

$$L^p(\mathbb{R}) \supset W_p^1(\mathbb{R}) \supset W_p^2(\mathbb{R}) \supset \dots .$$

The gaps between the spaces of such discrete Sobolev scale can be filled by the so-called *spectral approach*, i.e. by using the Fourier transform to describe smoothness. We denote by $\mathcal{F}f$ the Fourier transform of f given by

$$\mathcal{F}f(\omega) = (2\pi)^{-1/2} \int_{\mathbb{R}} \exp(-ix\omega) \cdot f(x) dx \qquad \omega \in \mathbb{R} \tag{3.1}$$

and by $\mathcal{F}^{-1}$ its inverse, cf. [67, Chapter 1]. The Fourier transform of the derivative can be described by

$$\mathcal{F}(\mathrm{D}f)(\omega) = (i\omega)\mathcal{F}(f)(\omega).$$

The importance of the above identity can hardly be overestimated. As a consequence we get

$$\|f\|^2_{W^k_2(\mathbb{R})} = \|\mathcal{F}^{-1}((1+|\omega|^k) \cdot \mathcal{F}f)\|^2_{L^2(\mathbb{R})} \sim \|\mathcal{F}^{-1}((1+|\omega|^2)^{k/2} \cdot \mathcal{F}f)\|^2_{L^2(\mathbb{R})} \quad (3.2)$$

where $\sim$ indicates equivalent norms. This shows that the smoothness of a function can be measured via its spectral decomposition.

In the next step we replace the $L_2(\mathbb{R})$ norm in the right-hand side of (3.2) by a $L^p(\mathbb{R})$. By doing so we define the so-called fractional Sobolev spaces via

$$H^s_p(\mathbb{R}) := \{f \ : \ \|f\|_{H^s_p} := \|\mathcal{F}^{-1}((1+|\omega|^2)^{s/2} \cdot \mathcal{F}f)\|_{L^p(\Omega)} < \infty\}.$$

We notice that the range of the parameter s is not restricted to non-negative values.

The spaces $H^s_p(\mathbb{R})$ are generalizations of the Sobolev spaces W^k_p. In fact, cf. [67, p. 12], for $k \in \mathbb{N}$ (and $1 < p < \infty$)

$$W^k_p(\mathbb{R}) = H^k_p(\mathbb{R}). \quad (3.3)$$

Therefore, fractional Sobolev spaces extend the discrete Sobolev scale to a continuous scale, where even negative degrees of smoothness are allowed.

3.2 (Sequence) Besov spaces

The aim of this section is to introduce the Besov norms. We again restrict ourselves first to the one-dimensional case. The extension to multi-dimensional case is *not* straight forward, cf. e.g. [68], but we remark that all subsequent results of this chapter hold also in the multi-dimensional case.

Before we define the Besov scale, we give a short introduction to the idea of wavelets, cf. [68, Sec. 1.7]. For other excellent introductions to the field of wavelets cf. [39, 14, 47].

A multi-resolution analysis is a sequence $\{V_j\}_{j\geq0}$ of subspaces of $L_2(\mathbb{R})$ such that

1. $V_0 \subset V_1 \subset \ldots \subset V_j \subset V_{j+1} \subset \ldots$ with $\operatorname{span} \bigcup_{j=0}^{\infty} V_j = L_2(\mathbb{R})$,

2. $f \in V_0$ if, and only if, $f(x-k) \in V_0$ for any $k \in \mathbb{Z}$,

3. $f \in V_j$ if, and only if, $f(2^{-j}x) \in V_0$

4. there is a function $\varphi \in V_0$ called *scaling function*, such that

$$\{\varphi(x-k) \ : \ k \in \mathbb{Z}\}$$

is an orthonormal basis of V_0.

By W_j we denote the orthogonal complement of V_j in V_{j+1}, i.e.

$$V_{j+1} = V_j \oplus W_j \,.$$

Then, by the properties of the multi resolution analysis we have

$$L_2(\mathbb{R}) = V_0 \oplus \bigoplus_{j=0}^{\infty} W_j.$$

Further, there exists a function $\psi \in W_0$ called *wavelet*, such that

$$\{\psi(x - k) \,:\, k \in \mathbb{Z}\}$$

is an orthonormal basis in W_0. We always think of ψ as a well localized, smooth wave (hence the name *wave*-let) with vanishing mean value. Examples of wavelets and scaling functions generating a multi-resolution analysis are shown in Figure 3.1.

To unify the notation for scaling functions and wavelets we introduce the functions

$$\psi_{j,k}(x) := \begin{cases} \varphi(x - k) & j = 0, \\ 2^{(j-1)/2}\psi(2^{(j-1)}x - k) & j \geq 1, k \in \mathbb{Z}. \end{cases}$$

This definition may seem unusual to some readers, since $W_j = \operatorname{span}_k\{\psi_{j+1,k}\}$. However, the main result of this section will be the Theorem 3.2, which we will cite from [68]. Therefore, we decided to keep the notation of [68].

With our notation the functions

$$\{\psi_{j,k} \,:\, j \geq 0, k \in \mathbb{Z}\}$$

form an orthonormal basis of $L^2(\mathbb{R})$. In the above definition of functions $\psi_{j,k}$ the number j indicates the so-called *scale* and the number k the position in space of the wavelet.

The projection of a function on the space V_0 is usually a low-pass filter and the projections on W_j are a series of high-pass filters. Since the functions $\psi_{j,k}$ are assumed to be well localized, the local spectrum of a function f can be described through the wavelet coefficients

$$f_{j,k} := \langle f, \psi_{j,k}\rangle = \int_{\mathbb{R}} f(x) \cdot \psi_{j,k}(x)dx.$$

Since $\psi_{j,k}$ form an orthogonal basis, we get that all $f \in L^2(\mathbb{R})$ can be represented as

$$f = \sum_{j,k} \psi_{j,k} f_{j,k}$$

and due to Parseval identity

$$\|f\|_{L^2(\mathbb{R})}^2 = \|(\|(f_{j,k})_k\|_{\ell^2}^2)_j\|_{\ell^2}^2 = \sum_{j \geq 0}\sum_{k \in \mathbb{Z}} |f_{j,k}|^2. \tag{3.4}$$

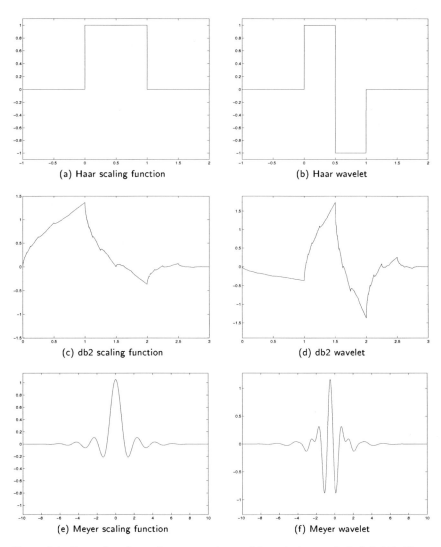

(a) Haar scaling function   (b) Haar wavelet

(c) db2 scaling function   (d) db2 wavelet

(e) Meyer scaling function   (f) Meyer wavelet

Figure 3.1: Examples of wavelets generating multi-resolution analysis (cf. [48, Chapter 7]). The db2 wavelet is an example of a compactly supported Daubechies wavelet, cf. [14].

We recall that in (3.2) the spaces $H_p^s(\mathbb{R})$ were introduced by multiplying the weights $(1 + |\omega|^2)^{s/2}$ on the spectrum on the function. Our understanding is that a function which has high oscillatory parts is not smooth. Therefore, the weight $(1 + |\omega|^2)^{s/2}$ ensures that the function does not contain too much of such high oscillatory parts.

Hence, it is clear that to get a smoothness measure with means of wavelet coefficients an appropriate weight on different scales has to be introduced. In analogy to the Sobolev case the weights should be increasing in j and s and constant in k. In the definition of the fractional Sobolev spaces $H_p^s(\mathbb{R})$ we also replaced the L^2 norm from (3.2) by the L^p norm. We can mimic this also in (3.4) and replace the norm with respect to the space index k by a ℓ^p norm, and the norm with respect to the scale index j by a ℓ^q norm. As it turns out, the appropriate weights are given by the numbers

$$2^{jsp} 2^{j(p-2)/2}.$$

If the wavelet is chosen smooth enough - for the exact conditions cf. [24, Chapter 7] - then the above considerations lead us directly to the definition of the so-called Besov spaces

$$B_{p,q}^s(\mathbb{R}) = \{f \ : \ \|f\|_{B_{p,q}^s(\mathbb{R})} := \|(f_{j,k})\|_{b_{p,q}^s} < \infty\}. \tag{3.5}$$

where

$$\|(f_{j,k})\|_{b_{p,q}^s} := \left(\sum_{j \geq 0} \left(\sum_{k \in \mathbb{Z}} 2^{jsp} 2^{j(p-2)/2} |f_{j,k}|^p \right)^{q/p} \right)^{1/q} \tag{3.6}$$

with the usual modifications for the case $p = \infty$ or $q = \infty$, cf. [68, Section 1.7], [24, Chapter 7] or [13, Chapter 3].

We remark that the definition of the spaces $B_{p,q}^s(\mathbb{R})$ is sensible for all $0 < p, q \leq \infty$ and $s \in \mathbb{R}$. En passant we can also define the sequence Besov spaces $b_{p,q}^s$ by

$$b_{p,q}^s = \{(f_{j,k}) \ : \ \|(f_{j,k})\|_{b_{p,q}^s} < \infty\}.$$

For the special value of $p = 2$ and $q = 2$ the spaces $B_{p,q}^s$ are closely connected to the Sobolev spaces $H_p^s(\mathbb{R})$.

Theorem 3.1. *[67, p. 12] Let $k \in \mathbb{N} \cup \{0\}$, then*

$$\|f\|_{W_2^k(\mathbb{R})} \sim \|f\|_{H_2^k(\mathbb{R})} \sim \|f\|_{B_{2,2}^k(\mathbb{R})}.$$

Let $s > 0$, then

$$\|f\|_{H_2^s(\mathbb{R})} \sim \|f\|_{B_{2,2}^s(\mathbb{R})}.$$

Further, Besov spaces are Banach spaces for $p, q \geq 1$ and by definition we have

$$f \in B_{p,q}^s(\mathbb{R}) \quad \Longrightarrow \quad (f_{j,k}) \in b_{p,q}^s.$$

But is every sequence in $b_{p,q}^s$ also the sequence of wavelets coefficients for some function in $B_{p,q}^s(\mathbb{R})$, i.e.

$$(\lambda_{j,k}) \in b_{p,q}^s \quad \Longrightarrow \quad \left(\sum_{j \geq 0} \sum_{k \in \mathbb{Z}} \lambda_{j,k} \psi_{j,k} \right) \in B_{p,q}^s(\mathbb{R})? \qquad (3.7)$$

Theorem 3.2. *Let $0 < p, q < \infty$ and $s \in \mathbb{R}$ and ψ be a sufficiently smooth wavelet (for exact conditions see [68, Theorem 1.64]) then the implication (3.7) is true with unconditional convergence being in the sense of $B_{p,q}^s(\mathbb{R})$. The wavelet decomposition $f \mapsto (f_{j,k})$ is an isomorphic map of $B_{p,q}^s(\mathbb{R})$ onto $b_{p,q}^s$.*

By the above theorem we do not loose any important properties if we consider only the spaces $b_{p,q}^s$ instead of $B_{p,q}^s(\mathbb{R})$. We think that in the light of applications sequence Besov spaces are more natural to be considered than the function Besov spaces, since the algorithms are usually applied to the wavelet-coefficients and not to the functions. Hence, in what follows we will focus on the sequence spaces $b_{p,q}^s$.

Therefore, the remaining part of the chapter will be devoted to the proof that for

$$1 < p, q < \infty$$

the spaces

$$b_{p,q}^s \quad \text{are} \quad \max\{2, p, q\}\text{-convex} \quad \text{and} \quad \min\{2, p, q\}\text{-smooth},$$

cf. Definitions 2.28 and 2.29. We will do this in a slightly more general setting. First, we will restate some well-known results on smoothness and convexity of sequence spaces ℓ^p. Then, we will introduce the weighted spaces ℓ_w^p equipped with the norm

$$\|(u_k)\|_{\ell_w^p} = \left(\sum_k w_k |u_k|^p \right)^{1/p}$$

and show that these spaces inherit all geometrical properties of the usual ℓ^p spaces as long as $w_k > 0$ for all k. Finally, we will introduce the spaces $\ell_w^{p,q}$

$$\|(u_{j,k})\|_{\ell_w^{p,q}} = \left(\sum_j \left(\sum_k w_{j,k} |u_{j,k}|^p \right)^{q/p} \right)^{1/q}$$

and show that these spaces are $\max\{2, p, q\}$-convex and $\min\{2, p, q\}$-smooth. The spaces $b_{p,q}^s$ are a special case of the spaces $\ell_w^{p,q}$ and have therefore the same geometrical properties.

3.3   Convexity and smoothness of ℓ^p spaces

For sequence spaces ℓ^p the claims of the following two theorems have been proven in [70] and [72].

Theorem 3.3. *For $1 < p < \infty$ we have*

$$\frac{1}{c}\|x - y\|_{\ell^p}^c \geq \frac{1}{c}\|x\|_{\ell^p}^c - \langle J_c^{\ell^p}(x), y \rangle + \frac{c_p}{c}\|y\|_{\ell^p}^c$$

where

$$c = \max\{p, 2\}$$

and

$$c_p = \min\{(p - 1), 2^{2-p}\}.$$

Theorem 3.4. *For $1 < p < \infty$ we have*

$$\frac{1}{s}\|x - y\|_{\ell^p}^s \leq \frac{1}{s}\|x\|_{\ell^p}^s - \langle J_s^{\ell^p}(x), y \rangle + \frac{G_p}{s}\|y\|_{\ell^p}^s$$

where

$$s = \min\{p, 2\}$$

and

$$G_p = \max\{(p - 1), 2^{2-p}\}$$

Remark 3.5. *One notices that the claims of Theorems 3.3 and 3.4 are similar to the claim of Example 2.38. In fact the existence of the constants c_p and G_p was already known in [45]. However, to the best of author's knowledge it was Xu who, in [70] and [72], provided the values presented above.*

Further, we remark that the nomenclature chosen for c_p and G_p is different than the nomenclature of Definitions 2.28 and 2.29, since in this section we want to emphasis the dependency on p and not on the convexity index c resp. the smoothness index s.

3.4   Spaces ℓ_w^p

Next, we consider the weighted spaces ℓ_w^p with norm

$$\|(u_k)\|_{\ell_w^p} := \left(\sum_k w_k |u_k|^p \right)^{1/p},$$

where $1 < p < \infty$ and w_k are strictly positive, i.e.

$$w_k > 0 \qquad \text{for all } k.$$

We have

$$\|(u_k)\|_{\ell_w^p} = \|(w_k^{1/p} \cdot u_k)\|_{\ell^p}. \tag{3.8}$$

We will see that the above identity allows us to transfer all important geometric properties from ℓ^p spaces to ℓ_w^p spaces. We also see that the standard ℓ^p spaces are just a special case of weighted spaces ℓ_w^p for $w_k = 1$.

Together with the dual mapping

$$\langle u, v^* \rangle_{\ell_w^p \times (\ell_w^p)^*} := \sum_k u_k v_k^*$$

the dual space of ℓ_w^p can be identified with a sequence space equipped with the norm

$$\|(v_k^*)\|_{(\ell_w^p)^*} = \left(\sum_k w_k^{1-p^*} |v_k^*|^{p^*} \right)^{1/p^*} = \left(\sum_k |w_k^{-1/p} v_k^*|^{p^*} \right)^{1/p^*}.$$

The proof of this claim can be carried out along the lines of [18, p. 286]. However, as a trust-ensuring measure we remark that the choice of weights for the dual norm ensures the continuity of the dual mapping $\langle \cdot, \cdot \rangle_{\ell_w^p \times (\ell_w^p)^*}$, since due to the Hölder inequality

$$\langle u, v^* \rangle_{\ell_w^p \times (\ell_w^p)^*} = \sum_k u_k v_k^*$$
$$\leq \sum_k |w_k^{1/p} u_k| \cdot |w_k^{-1/p} v_k^*|$$
$$\leq \left(\sum_k w_k |u_k|^p \right)^{1/p} \cdot \left(\sum_k w_k^{-p^*/p} |v_k^*|^{p^*} \right)^{1/p^*} < \infty.$$

One can check that the mapping $j_r^{\ell_w^p}$ given by

$$\left(j_r^{\ell_w^p}(u) \right)_k := \|u\|_{\ell_w^p}^{r-p} \cdot w_k |u_k|^{p-1} \cdot \operatorname{sign}(u_k) \tag{3.9}$$

is an element of the duality mapping $J_r^{\ell_w^p}$. We will show that the spaces ℓ_w^p are smooth of power type, which ensures that $j_r^{\ell_w^p}$ is actually the only element of $J_r^{\ell_w^p}$.

Corollary 3.6. *Theorems 3.3 and 3.4 remain valid if the ℓ^p norms are replaced by ℓ_w^p norms.*

Proof. Assume that $u, v \in \ell_w^p$, then by (3.8) we have $(w_k^{1/p} \cdot u_k), (w_k^{1/p} \cdot v_k) \in \ell^p$. By Theorem 3.4 we have for the same s and G_p as in Theorem 3.4 that

$$\frac{1}{s} \|(w_k^{1/p} \cdot (u_k - v_k))\|_{\ell^p}^s$$
$$\leq \frac{1}{s} \|(w_k^{1/p} \cdot u_k)\|_{\ell^p}^s - \langle j_s^{\ell^p}((w_k^{1/p} \cdot u_k)), (w_k^{1/p} \cdot v_k) \rangle + \frac{G_p}{s} \|w_k^{1/p} \cdot v_k\|_{\ell^p}^s.$$

Together with (3.8) and

$$\langle j_s^{\ell^p}((w_k^{1/p} \cdot u_k)), (w_k^{1/p} \cdot v_k)\rangle_{(\ell^p)^* \times \ell^p} = \langle j_s^{\ell_w^p}(u), v\rangle_{(\ell_w^p)^* \times \ell_w^p},$$

we get that the inequality of Theorem 3.4 holds for the particular choice of the duality mapping as in (3.9).

Due to the Xu-Roach characterization of smoothness of power type in Theorem 2.33 we get that the duality mapping $J_s^{\ell_w^p}$ is single-valued. Hence, Theorem 3.4 holds for ℓ_w^p.

By the properties of the duality mapping, cf. Theorem 2.44, we get that the duality mapping $J_c^{\ell_w^p}$ is single-valued too. By the same argumentation as above the inequality of Theorem 3.3 holds if we replace ℓ^p by ℓ_w^p.   □

Hence, the spaces ℓ_w^p are smooth of power type and

$$\{j_r^{\ell_w^p}\} = J_r^{\ell_w^p}.$$

We also remark that - surprisingly - the above theorem holds for all strictly positive weights. In particular assumptions like $w_k \to 0$ or $w_k \to \infty$ do *not* have to be assumed, which is the main difference to the necessary assumptions of Daubechies, Defrise and De Mol [15].

3.5 Spaces $\ell_w^{p,q}$

In the next step we consider the spaces $\ell_w^{p,q}$ with the norm

$$\|(u_{j,k})\|_{\ell_w^{p,q}} := \left(\sum_j \left(\sum_k w_{j,k}|u_{j,k}|^p\right)^{q/p}\right)^{1/q}.$$

With definitions introduced in the last section we could also write

$$\|(u_{j,k})\|_{\ell_w^{p,q}} = \left\|\left(\left(\|((u_{j,k})_k)\|_{\ell_{w_j}^p}\right)_j\right)\right\|_{\ell^q},$$

with the sequence (w_j) of weight sequences defined via

$$w_j := ((w_{j,k})_k),$$

where by $((w_{j,k})_k)$ we mean for every fixed j the sequence (v_k) with $v_k = w_{j,k}$, i.e. the (sub-)sequence with respect to k. Analogously $\|((u_{j,k})_k)\|$ is a sequence of numbers with sequence index j.

We stress that p and q do *not* need to be dual exponents. We denote the dual exponents of p and q by p^* and q^*, i.e.

$$\frac{1}{p} + \frac{1}{p^*} = 1 \qquad \frac{1}{q} + \frac{1}{q^*} = 1.$$

Together with the dual mapping

$$\langle (v_{j,k}^*), (u_{j,k}) \rangle_{(\ell_w^{p,q})^* \times \ell_w^{p,q}} := \sum_j \sum_k v_{j,k}^* u_{j,k}$$

the dual space of $\ell_w^{p,q}$ can be identified with a sequence space equipped with the norm

$$\|v^*\|_{(\ell_w^{p,q})^*} := \left\| \left(\left\| ((v_{j,k}^*)_k) \right\|_{(\ell_{w_j}^p)^*} \right)_j \right\|_{\ell^{q^*}} = \left(\sum_j \left(\sum_k w_{j,k}^{1-p^*} |v_{j,k}^*|^{p^*} \right)^{q^*/p^*} \right)^{1/q^*}.$$
$$(3.10)$$

This ensures the continuity of the dual mapping, since due to Hölder inequality

$$\langle (v_{j,k}^*), (u_{j,k}) \rangle_{(\ell_w^{p,q})^* \times \ell_w^{p,q}} = \sum_j \sum_k v_{j,k}^* u_{j,k}$$

$$\leq \sum_j \left\| ((u_{j,k})_k) \right\|_{\ell_{w_j}^p} \cdot \left\| ((v_{j,k}^*)_k) \right\|_{(\ell_{w_j}^p)^*}$$

$$\leq \left\| \left(\left\| (u_{j,k})_k \right\|_{\ell_{w_j}^p} \right)_j \right\|_{\ell^q} \cdot \left\| \left(\left\| (v_{j,k}^*)_k \right\|_{(\ell_{w_j}^p)^*} \right)_j \right\|_{\ell^{q^*}}.$$

One can check that the mapping

$$\left(j_r^{\ell_w^{p,q}}((u_{j,k})) \right)_{a,b} := \|(u_{j,k})\|_{\ell_w^{p,q}}^{r-q} \cdot \|((u_{a,k})_k)\|_{\ell_{w_a}^p}^{q-p} \cdot w_{a,b} |u_{a,b}|^{p-1} \operatorname{sign}(u_{a,b})$$

is an element of the duality mapping $J_r^{\ell_w^{p,q}}$. By $\left(j_r^{\ell_w^{p,q}}((u_{j,k})) \right)_{a,b}$ we denote the a,b-th element of $j_r^{\ell_w^{p,q}}((u_{j,k}))$. In the next section we will show that the spaces $\ell_w^{p,q}$ are smooth of power type and therefore $j_r^{\ell_w^{p,q}}$ is the only element of $J_r^{\ell_w^{p,q}}$, i.e.

$$\{ j_r^{\ell_w^{p,q}} \} = J_r^{\ell_w^{p,q}}.$$

3.6   Convexity and smoothness of $\ell_w^{p,q}$

We already know that spaces ℓ^p and ℓ_w^p are $\min\{2,p\}$-smooth and $\max\{2,p\}$-convex. In this section, we will show that for

$$1 < p, q < \infty$$

the spaces $\ell_w^{p,q}$ are

$$\min\{2,p,q\}\text{-smooth}$$

and

$$\max\{2,p,q\}\text{-convex.}$$

Theorem 3.7. *For $1 < p,q < \infty$ the spaces $\ell_w^{p,q}$ are $\min\{2,p,q\}$-smooth.*

Proof. First, we consider the case $q \le \min\{p,2\}$. The spaces $\ell_{w_j}^p$ are $\min\{p,2\}$-smooth. Due to $q \le \min\{p,2\}$ they are also q-smooth by Corollary 2.34. Therefore, by Corollary 3.6 and Theorem 2.33 there exists a constant C such that

$$\|j_q^{\ell_{w_j}^p}(((u_{j,k})_k)) - j_q^{\ell_{w_j}^p}(((v_{j,k})_k))\|_{(\ell_{w_j}^p)^*} \le C\|((u_{j,k} - v_{j,k})_k)\|_{\ell_{w_j}^p}^{q-1}.$$

where the value of C is independent of the weights w_j. Since

$$(j_q^{\ell_w^{p,q}}((u_{j,k})))_{a,b} = (j_q^{\ell_{w_a}^p}(((u_{a,k})_k)))_b$$

we have

$$\|j_q^{\ell_w^{p,q}}((u_{j,k})) - j_q^{\ell_w^{p,q}}((v_{j,k}))\|_{(\ell_w^{p,q})^*}^{q^*} = \sum_j \|j_q^{\ell_{w_j}^p}(((u_{j,k})_k)) - j_q^{\ell_{w_j}^p}(((v_{j,k})_k))\|_{(\ell_{w_j}^p)^*}^{q^*}$$

$$\le C\sum_j \|((u_{j,k} - v_{j,k})_k)\|_{\ell_{w_j}^p}^{(q-1)q^*}$$

$$= C\|(u_{j,k} - v_{j,k})\|_{\ell_w^{p,q}}^q.$$

Then, after taking the q^*th root we see that $\ell_w^{p,q}$ are q-smooth by the Xu-Roach characterization of smoothness of power type in Theorem 2.33.

Next, we consider the case $q > \min\{p,2\}$. The spaces $\ell_{w_j}^p$ are $\min\{p,2\}$-smooth, without loss of generality we assume that $p = \min\{p,2\}$. Then, by the Xu-Roach characterization of smoothness of power type in Theorem 2.33 there exists a constant C - which as before is independent of the weights w_j - such that

$$\|j_q^{\ell_{w_j}^p}(((u_{j,k})_k)) - j_q^{\ell_{w_j}^p}(((v_{j,k})_k))\|_{(\ell_{w_j}^p)^*}$$

$$\le C\left(\max\{\|((u_{j,k})_k)\|_{\ell_{w_j}^p}, \|((v_{j,k})_k)\|_{\ell_{w_j}^p}\}\right)^{q-p}\|((u_{j,k} - v_{j,k})_k)\|_{\ell_{w_j}^p}^{p-1}.$$

Therefore, we have

$$\|j_q^{\ell_w^{p,q}}((u_{j,k})) - j_q^{\ell_w^{p,q}}((v_{j,k}))\|_{(\ell_w^{p,q})^*}^{q^*}$$

$$= \sum_j \|j_q^{\ell_{w_j}^p}(((u_{j,k})_k)) - j_q^{\ell_{w_j}^p}(((v_{j,k})_k))\|_{(\ell_{w_j}^p)^*}^{q^*}$$

$$\le C\sum_j \left(\left(\max\{\|((u_{j,k})_k)\|_{\ell_{w_j}^p}, \|((v_{j,k})_k)\|_{\ell_{w_j}^p}\}\right)^{q-p}\|((u_{j,k} - v_{j,k})_k)\|_{\ell_{w_j}^p}^{p-1}\right)^{q^*}$$

Since $q - p \geq 0$ and $\max\{|a|, |b|\}^\alpha = \max\{|a|^\alpha, |b|^\alpha\}$ for $\alpha \geq 0$ we get

$$\|j_q^{\ell_w^{p,q}}(u_{j,k}) - j_q^{\ell_w^{p,q}}(v_{j,k})\|_{(\ell_w^{p,q})^*}^{q^*}$$

$$\leq C \sum_j \left(\left(\max\{\|((u_{j,k})_k)\|_{\ell_{w_j}^p}^{q-p}, \|((v_{j,k})_k)\|_{\ell_{w_j}^p}^{q-p}\} \right) \|((u_{j,k} - v_{j,k})_k)\|_{\ell_{w_j}^p}^{p-1} \right)^{q^*}$$

We have

$$\sum_j \|((u_{j,k})_k)\|_{\ell_{w_j}^p}^{(q-p)q^*} \|((u_{j,k} - v_{j,k})_k)\|_{\ell_{w_j}^p}^{(p-1)q^*}$$

$$= \sum_j \left(\|((u_{j,k})_k)\|_{\ell_{w_j}^p}^q \right)^{\frac{(q-p)q^*}{q}} \left(\|((u_{j,k} - v_{j,k})_k)\|_{\ell_{w_j}^p}^q \right)^{\frac{(p-1)q^*}{q}}$$

$$\leq \left(\sum_j \|((u_{j,k})_k)\|_{\ell_{w_j}^p}^q \right)^{\frac{(q-p)q^*}{q}} \left(\sum_j \|((u_{j,k} - v_{j,k})_k)\|_{\ell_{w_j}^p}^q \right)^{\frac{(p-1)q^*}{q}}$$

$$= \|(u_{j,k})\|_{\ell_w^{p,q}}^{(q-p)q^*} \|(u_{j,k}) - (v_{j,k})\|_{\ell_w^{p,q}}^{(p-1)q^*}.$$

Above, we used the Hölder-inequality

$$\sum_j |x_j y_j| \leq (\sum_j |x_j|^P)^{1/P} \cdot (\sum_j |y_j|^{P^*})^{1/P^*}.$$

with parameters

$$1/P = \frac{(q-p)q^*}{q} \qquad\qquad 1/P^* = \frac{(p-1)q^*}{q}$$

$$x_j = \left(\|((u_{j,k})_k)\|_{\ell_{w_j}^p}^q \right)^{\frac{(q-p)q^*}{q}} \qquad y_j = \left(\|((u_{j,k} - v_{j,k})_k)\|_{\ell_{w_j}^p}^q \right)^{\frac{(p-1)q^*}{q}}.$$

We remark that $P^* > 1$ since $\min\{p, 2\} < q$. We use the same estimation for the $\|((v_{j,k})_k)\|$ term and arrive at

$$\|j_q^{\ell_w^{p,q}}((u_{j,k})) - j_q^{\ell_w^{p,q}}((v_{j,k}))\|_{(\ell_w^{p,q})^*}^{q^*}$$

$$\leq C \left(\max\{\|(u_{j,k})\|_{\ell_w^{p,q}}, \|(v_{j,k})\|_{\ell_w^{p,q}}\} \right)^{(q-p)q^*} \|(u_{j,k}) - (v_{j,k})\|_{\ell_w^{p,q}}^{(p-1)q^*}$$

After taking the q^*th root we see see that $\ell_w^{p,q}$ are p-smooth by the Xu-Roach characterization of smoothness of power type in Theorem 2.33. $\qquad\square$

Theorem 3.8. *For $1 < p, q < \infty$ the spaces $\ell_w^{p,q}$ are $\max\{2, p, q\}$-convex.*

Proof. By (3.10) the space $\ell_{w^*}^{p^*,q^*}$ with weights

$$w_{j,k}^* = w_{j,k}^{1-p^*}$$

is the dual space $(\ell_w^{p,q})^*$. By Theorem 3.7 the space $\ell_{w^*}^{p^*,q^*}$ is $\min\{2,p^*,q^*\}$-smooth. Therefore, by Theorem 2.42 also reflexive and $(\ell_{w^*}^{p^*,q^*})^* = (\ell_w^{p,q})^{**} = \ell_w^{p,q}$. Hence, by Theorem 2.43 the dual space of $\ell_{w^*}^{p^*,q^*}$, i.e. $\ell_w^{p,q}$ is $(\min\{2,p^*,q^*\})^*$-convex. Finally, we get

$$(\min\{2,p^*,q^*\})^* = \max\{2,p,q\},$$

which proves the claim.   □

3.7   Consequences to sequence Besov spaces

In this section, we will show how the results of the previous sections apply to the Besov sequence spaces. The main aim of this section is to provide a reference for the geometrical facts concerning Besov spaces.

We recall that due to Theorem 3.2 there is a tight connection between the function Besov spaces $B_{p,q}^s(\mathbb{R})$ and the sequence Besov spaces $b_{p,q}^s$ equipped with the norm

$$\|(f_{j,k})\|_{b_{p,q}^s} = \left(\sum_j \left(\sum_k 2^{jsp} 2^{j(p-2)/2} |f_{j,k}|^p \right)^{q/p} \right)^{1/p}.$$

The above norm is a special case of the weighted norms considered in Sections 3.5 and 3.6.

Since

$$(jsp + j(p-2)/2) \cdot (1 - p^*) = -jsp^* + j(p^*-2)/2$$

the dual space $(b_{p,q}^s)^*$ can be identified with a sequence space equipped with the norm

$$\|(g_{j,k}^*)\|_{(b_{p,q}^s)^*} = \left(\sum_j \left(\sum_k 2^{-jsp^*} 2^{j(p^*-2)/2} |f_{j,k}|^{p^*} \right)^{q^*/p^*} \right)^{1/p^*}$$

and the dual mapping

$$\langle (f_{j,k}), (g_{j,k}^*) \rangle_{b_{p,q}^s \times (b_{p,q}^s)^*} = \sum_{j,k} f_{j,k} g_{j,k}^*.$$

Hence

$$(b_{p,q}^s)^* = b_{p^*,q^*}^{-s}.$$

Due to Theorems 3.7 and 3.8 we know that the spaces $b^s_{p,q}$ are

$$\max\{2, p, q\}\text{-convex} \quad \text{and} \quad \min\{2, p, q\}\text{-smooth}$$

for $1 < p, q < \infty$ and $s \in \mathbb{R}$.

In most applications the parameters s and p are given by the underlying problem. By contrast the parameter q can be chosen more or less freely and is therefore known as *fine-tuning parameter* (cf. [17]). From the geometrical point of view the best choice is to set

$$q = p$$

since in this way one never looses any convexity or smoothness properties. This means that from purely geometrical point of view the sequence Besov spaces $b^s_{p,p}$ are the most well-behaved and therefore the most interesting ones. The norm of $b^s_{p,p}$ is given by

$$\|(f_{j,k})\|_{b^s_{p,p}} = \left(\sum_j \sum_k 2^{jsp} 2^{j(p-2)/2} |f_{j,k}|^p \right)^{1/p}.$$

Further, the space $b^s_{p,p}$ is $\max(2, p)$-convex and $\min(2, p)$-smooth and for the dual space we have $(b_{p,p})^* = b^{-s}_{p^*,p^*}$. The duality mapping is given by

$$\left(J^{b^s_{p,p}}_r((f_{j,k})) \right)_{a,b} = \|(f_{j,k})\|^{r-p}_{b^s_{p,p}} \cdot 2^{asp} 2^{a(p-2)/2} \cdot |f_{a,b}|^{p-1} \cdot \operatorname{sign}(f_{a,b})$$

and the inequalities

$$\frac{1}{c} \|(f_{j,k}) - (g_{j,k})\|^c_{b^s_{p,p}} \geq \frac{1}{c} \|(f_{j,k})\|^c_{b^s_{p,p}} - \langle J^{b^s_{p,p}}_c((f_{j,k})), (g_{j,k}) \rangle + \frac{c_p}{c} \|(g_{j,k})\|^c_{b^s_{p,p}},$$

$$\frac{1}{\sigma} \|(f_{j,k}) - (g_{j,k})\|^\sigma_{b^s_{p,p}} \leq \frac{1}{\sigma} \|(f_{j,k})\|^\sigma_{b^s_{p,p}} - \langle J^{b^s_{p,p}}_\sigma((f_{j,k})), (g_{j,k}) \rangle + \frac{G_p}{\sigma} \|(g_{j,k})\|^\sigma_{b^s_{p,p}}$$

hold with

$$c = \max\{p, 2\} \qquad c_p = \min\{(p-1), 2^{2-p}\} \quad \text{and}$$
$$\sigma = \min\{p, 2\} \qquad G_p = \max\{(p-1), 2^{2-p}\}.$$

Remark 3.9. *Similar inequalities can be proven for all Besov spaces $b^s_{p,q}$ with $1 < p, q < \infty$ and $s \in \mathbb{R}$. The tricky part is to find a good estimate on the related constants $c_{p,q}$ and $G_{p,q}$. Rough estimates may be obtained by the methods used in [71, 70, 72].*

4

Tikhonov regularization

We return to the question of regularizing the operator equation

$$Ax = y.$$

First, we recall some well-known facts [23] from the Hilbert space setting, i.e. the case that $A : X \to Y$ with X and Y being Hilbert spaces. The minimizers x_α^δ of the Tikhonov functional T_α defined via

$$T_\alpha(x) = \tfrac{1}{2}\|Ax - y\|^2 + \alpha \tfrac{1}{2}\|x\|^2$$

can be used to regularize the operator equation.

First, we recall that x_α^δ form a regularizing scheme — i.e. with $x^\dagger$ the minimum norm solution of $Ax = y$ we have $x_\alpha^\delta \to x^\dagger$ as $\delta \to 0$ — if

$$\alpha(\delta) \to 0 \qquad \text{and} \qquad \frac{\delta^2}{\alpha(\delta)} \to 0 \qquad \text{as} \quad \delta \to 0.$$

However, due to the Lethargy Theorem of Regularization (cf. [57], [23, Proposition 3.11]) the convergence of x_α^δ to $x^\dagger$ can be arbitrarily slow. Therefore, additional conditions have to be implied on $x^\dagger$ to achieve convergence rates results. One can show that for $x^\dagger$ with source condition

$$x^\dagger \in \mathcal{R}((A^*A)^{\mu/2}) \tag{4.1}$$

we have

$$\|x_\alpha^\delta - x^\dagger\| \leq \mathcal{C} \cdot \delta^{\mu/(\mu+1)} \tag{4.2}$$

if α, as a function of δ, is chosen appropriately and $\mu \leq 2$. For source conditions with $\mu > 2$ the rate is the same as for $\mu = 2$. Therefore, the best possible rate which can be achieved by Tikhonov regularization is given by $\|x_\alpha^\delta - x^\dagger\| \leq C \cdot \delta^{2/3}$ or

$$\tfrac{1}{2}\|x_\alpha^\delta - x^\dagger\|^2 \leq C \cdot \delta^{4/3}. \tag{4.3}$$

Next, we recall that one can also choose α in such a way that for some appropriately chosen τ we have

$$\|Ax_\alpha^\delta - y^\delta\| = \tau\delta.$$

Such choice is known as Morozov's discrepancy principle. It is well-known that the convergence rate (4.2) can only be obtained for $\mu \leq 1$. For $\mu > 1$ the same rate as for $\mu = 1$ holds, namely $\|x_\alpha^\delta - x^\dagger\| \leq C \cdot \delta^{1/2}$ or

$$\tfrac{1}{2}\|x_\alpha^\delta - x^\dagger\|^2 \leq C \cdot \delta.$$

To achieve better rates some other parameter choice rule has to be applied. In [22] it was shown that the best possible rate (4.3) may be achieved under the source condition (4.1) with $\mu = 2$ if α is chosen such that

$$\|A^*Ax_\alpha^\delta - A^*y^\delta\|^2 = \delta^2\alpha^{-1}.$$

The above discrepancy principle is known as Engl's discrepancy principle.

Finally, we recall that the minimizers of the Tikhonov functional are computed by solving the normal equation

$$\nabla T_\alpha(x_\alpha^\delta) = A^*(Ax_\alpha^\delta - y^\delta) + \alpha x_\alpha^\delta = 0.$$

However, to solve the above equation - especially for very high dimensional discretizations of the operator A and A^* - one has to use an iterative solver, e.g. the steepest descent method

$$x_{n+1} = x_n - \mu_n \nabla T_\alpha(x_n). \tag{4.4}$$

The aim of this chapter is to show how and to what extent the above results can be translated to the setting of Banach spaces, i.e. the setting where $A : X \to Y$ with X and Y being Banach spaces. For the Tikhonov functional

$$T_\alpha(x) = \tfrac{1}{p}\|Ax - y\|_Y^p + \alpha \cdot \tfrac{1}{q}\|x\|_X^q, \qquad p, q > 1,$$

we will consider the following points:

1. (Regularization properties) First, we will show that the minimizers

$$x_\alpha^\delta \in \arg\min_{x \in X} T_\alpha(x)$$

exist and are unique if X is convex of power type. Further, we will show that $x_\alpha^\delta \to x^\dagger$ as long as

$$\alpha(\delta) \to 0 \qquad \text{and} \qquad \frac{\delta^p}{\alpha(\delta)} \to 0 \qquad \text{as} \quad \delta \to 0.$$

2. (Source conditions) Second, we will try to generalize two special cases of the range source conditions (4.1). To the best of author's knowledge, this is particularly straight forward for $\mu = 1$ and $\mu = 2$, i.e. for the source conditions $x^\dagger \in \mathcal{R}((A^*A)^{1/2}) = \mathcal{R}(A^*)$ and $x^\dagger \in \mathcal{R}(A^*A)$. Then, the source conditions in the Banach space setting read as

$$j_q(x^\dagger) = A^*\omega \qquad \text{and} \qquad j_q(x^\dagger) = A^*j_p(A\omega).$$

We call the first source condition low order source condition and the second one high order source condition.

3. (Parameter choice) Next, using appropriate source conditions and parameter choice rules, we will show convergence rates of the form

$$D_{j_q}(x^\dagger, x_\alpha^\delta) \le C \cdot \delta^\kappa,$$

i.e. convergence rates with respect of the Bregman distance. Further, we will show that Morozov's discrepancy principle together with the low order source condition yields the convergence rate

$$D_{j_q}(x^\dagger, x_\alpha^\delta) \le C \cdot \delta.$$

Together with the results of the Hilbert space setting, it seems that no better convergence rates can be achieved, even if stronger source conditions are assumed. To achieve better rates, we will develop and employ a version of the discrepancy principle of Engl.

4. (Minimization schemes) Finally, to apply the results described above, we will need algorithms which are able to minimize the Tikhonov functional. We will therefore introduce two generalizations for the steepest descent (4.4), which are given by

$$x_{n+1} = x_n - \mu_n J_{q^*}^{X^*}(\psi_n) \quad \text{with} \quad \psi_n \in \nabla T_\alpha(x_n)$$

and

$$x_{n+1}^* = x_n^* - \mu_n \psi_n \quad \text{with} \quad \psi_n \in \partial T_\alpha(x_n)$$
$$x_{n+1} = J_{q^*}^{X^*}(x_{n+1}^*).$$

We will show strong convergence rates for both algorithms under appropriate assumptions on X and Y.

As last part of the introduction to this chapter, we want to mention the notation which will hold throughout this chapter.

Remark 4.1. *If not stated otherwise:*

1. *We denote by $x^\dagger$ the minimum norm solution of $Ax = y$, cf. Definition 2.53.*

2. *The operator $A : X \to Y$ is continuous and linear and the spaces X and Y are Banach spaces.*

3. *For $p > 1$ and $q > 1$ the Tikhonov functional T_α is given by*

$$\mathrm{T}_\alpha(x) := \tfrac{1}{p}\|Ax - y^\delta\|_Y^p + \alpha \cdot \tfrac{1}{q}\|x\|_X^q.$$

4. *By x_α^δ we denote a minimizer of the Tikhonov functional T_α, i.e.*

$$x_\alpha^\delta \in \operatorname{argmin}_{x \in X} \mathrm{T}_\alpha(x).$$

5. *We denote by y^δ the noisy version of y with noise-level δ, i.e.*

$$\|y^\delta - y\| \le \delta.$$

4.1 Regularization properties of Tikhonov functionals

First, we will show that for spaces X convex of power type the minimizers x_α^δ of the Tikhonov functional $\mathrm{T}_\alpha(x) = \frac{1}{p}\|Ax - y^\delta\|^p + \alpha \cdot \frac{1}{q}\|x\|^q$ can be used to regularize the problem $Ax = y$.

Theorem 4.2. *Let Y be a Banach space, X a Banach space convex of power type, $A : X \to Y$ linear and continuous and x_α^δ the minimizer of the Tikhonov functional $\mathrm{T}_\alpha(x) = \frac{1}{p}\|Ax - y^\delta\|_Y^p + \alpha \cdot \frac{1}{q}\|x\|_X^q$ with $p, q > 1$. Assume that there exists a solution of $Ax = y$. Moreover, assume that $\alpha(\delta, y^\delta)$ satisfies for all y^δ with $\|y - y^\delta\| \le \delta$*

$$\alpha(\delta, y^\delta) \to 0 \quad \text{and} \quad \frac{\delta^p}{\alpha(\delta, y^\delta)} \to 0, \quad \text{as} \quad \delta \to 0.$$

Then, for every sequence (y_n) with $\|y - y_n\| \le \delta_n$ and $\delta_n \to 0$ the sequence (x_n) of unique minimizers of $\frac{1}{p}\|Ax - y_n\|_Y^p + \alpha(\delta_n, y_n) \cdot \frac{1}{q}\|x\|_X^q$ converges (strongly) to the unique norm-minimizing solution $x^\dagger$ of $Ax = y$.

Proof. The claim is a consequence of the results of [55, Section 3.2] together with the convexity of power type of X. For a detailed proof cf. Appendix A. $\quad\square$

Remark 4.3. *The above theorem holds if Y is an arbitrary Banach space and X is a sequence space ℓ^p, a Lebesgue space $L^p(\Omega)$, a Sobolev space $W_k^p(\Omega)$, or a (sequence) Besov space $B_{p,q}^s(\mathbb{R}^d)$ resp. $b_{p,q}^s$ with*

$$1 < p, q < \infty,$$

cf. Example 2.38 and Subsection 3.7.

4.2   Source conditions

Let x_α^δ be a minimizer of the Tikhonov functional

$$\mathrm{T}_\alpha(x): \quad = \quad \tfrac{1}{p}\|Ax - y^\delta\|^p + \alpha \cdot \tfrac{1}{q}\|x\|^q,$$

i.e.

$$x_\alpha^\delta \in \mathrm{argmin}_{x \in X}\, \mathrm{T}_\alpha(x).$$

As a consequence of the results in the last section, we know that the minimizers x_α^δ converge to the minimal-norm solution $x^\dagger$ of $Ax = y$, as long as the parameter $\alpha = \alpha(\delta)$ is chosen appropriately.

However, due to the Lethargy Theorem of Regularization (cf. [57], [23, Proposition 3.11]) the convergence can be arbitrarily slow. Therefore, some additional conditions have to be implied to achieve convergence rates results. Since usually such additional conditions are implied to the minimum norm solution $x^\dagger$, i.e. the source of the data y, they are referred to as *source conditions*.

Throughout this section, we assume that X and Y are given such that the notions of minimizer x_α^δ of T_α and of minimum norm solution $x^\dagger$ are sensible, i.e. x_α^δ and $x^\dagger$ exist. However, we do not make any explicit assumptions on the spaces X and Y. An example of a sensible setting for X and Y was presented in the last section.

4.2.1   Low order source condition

In the Hilbert space setting, i.e. for X and Y being Hilbert spaces, the probably most used source conditions are the so-called *range source conditions*, which are given by $x^\dagger \in \mathcal{R}((A^*A)^{\mu/2})$ for some $\mu > 0$.

It turns out that in the case of Tikhonov regularization the easiest way to translate the concept of range source conditions into the realm of Banach spaces is to consider $\mu = 1$, since for this particular value of μ we have $\mathcal{R}(A^*A)^{\mu/2} = \mathcal{R}(A^*A)^{1/2} = \mathcal{R}(A^*)$. Therefore, for $\mu = 1$ one can rewrite the range source condition as

$$x^\dagger = A^*\omega, \tag{4.5}$$

for some $\omega \in Y^*$. In [10] the authors propose to generalize the above source condition to the setting of Banach spaces by

$$j_q(x^\dagger) = A^*\omega, \tag{4.6}$$

for some $j_q \in J_q$ and $\omega \in Y^*$. Then, the following result holds:

Theorem 4.4. *Let $x^\dagger$ be a minimum norm solution of $Ax = y$ such that the range source condition*

$$j_q(x^\dagger) = A^*\omega$$

holds for some $j_q \in J_q$ *and* $\omega \in Y^*$. *Let* x_α^δ *be a minimizer of the Tikhonov functional* $\mathrm{T}_\alpha(x) := \frac{1}{p}\|Ax - y^\delta\|^p + \alpha \cdot \frac{1}{q}\|x\|^q$, *where* $p, q > 1$ *and* $\|y - y^\delta\| \leq \delta$. *Then*

$$D_{j_q}(x^\dagger, x_\alpha^\delta) \leq \alpha^{-1} \cdot \frac{1}{p}(\delta^p - \|Ax_\alpha^\delta - y^\delta\|^p) + \|\omega\|(\|Ax_\alpha^\delta - y^\delta\| + \delta)$$

and

$$D_{j_q}(x^\dagger, x_\alpha^\delta) \leq \frac{1}{p^*}\|\omega\|^{p^*} \cdot \alpha^{p^*-1} + \frac{1}{p}\alpha^{-1} \cdot \delta^p + \|\omega\|\delta.$$

Proof. Since x_α^δ minimizes the Tikhonov functional, we have

$$\begin{aligned}
\frac{1}{p}\|Ax_\alpha^\delta &- y^\delta\|^p + \alpha D_{j_q}(x^\dagger, x_\alpha^\delta)\\
&= \frac{1}{p}\|Ax_\alpha^\delta - y^\delta\|^p + \alpha\frac{1}{q}\|x_\alpha^\delta\|^q - \alpha\frac{1}{q}\|x^\dagger\|^q - \alpha\langle j_q(x^\dagger), x_\alpha^\delta - x^\dagger\rangle\\
&\leq \frac{1}{p}\|Ax^\dagger - y^\delta\|^p + \alpha\frac{1}{q}\|x^\dagger\|^q - \alpha\frac{1}{q}\|x^\dagger\|^q - \alpha\langle j_q(x^\dagger), x_\alpha^\delta - x^\dagger\rangle\\
&\leq \frac{1}{p}\delta^p - \alpha\langle j_q(x^\dagger), x_\alpha^\delta - x^\dagger\rangle.
\end{aligned}$$

Due to the source condition, we get

$$\begin{aligned}
-\alpha\langle j_q(x^\dagger), x_\alpha^\delta - x\rangle &= \alpha\langle -\omega, Ax_\alpha^\delta - Ax^\dagger\rangle \leq \alpha\|\omega\|\|Ax_\alpha^\delta - y\|\\
&\leq \alpha\|\omega\|\|Ax_\alpha^\delta - y^\delta\| + \alpha\|\omega\|\delta,
\end{aligned}$$

which proves the first claim. By Young's inequality, we get

$$\alpha\|\omega\|\|Ax_\alpha^\delta - y^\delta\| \leq \frac{1}{p}\|Ax_\alpha^\delta - y^\delta\|^p + \frac{1}{p^*}\|\omega\|^{p^*} \cdot \alpha^{p^*}.$$

And therefore

$$\begin{aligned}
\frac{1}{p}\|Ax_\alpha^\delta &- y^\delta\|^p + \alpha D_\xi(x^\dagger, x_\alpha^\delta)\\
&\leq \frac{1}{p}\|Ax_\alpha^\delta - y^\delta\|^p + \frac{1}{p^*}\|\omega\|^{p^*} \cdot \alpha^{p^*} + \frac{1}{p}\delta^p + \alpha\|\omega\|\delta,
\end{aligned}$$

which is equivalent to the second claim.   □

Remark 4.5. *The astonishing fact about the above theorem is therefore that no properties of the space* X *are being used. However, to translate the convergence with respect to the Bregman distance into strong convergence, one has to assume a convexity property on* X *like uniform convexity or convexity of power type.*

4.2.2 High order source condition

In Tikhonov regularization with Hilbert space norms, one can show that if the minimum norm solution $x^\dagger$ admits the source condition

$$x^\dagger = A^*A\omega \qquad \text{for some } \omega \in X,$$

then
$$\|x^\dagger - x_\alpha^\delta\| \leq C \cdot \delta^{2/3}$$

for an appropriate choice of α. Further, it is well-known that no faster convergence can be obtained, even if the minimal norm solution admits an even stronger source condition. In other words, for Tikhonov regularization the rate $\delta^{2/3}$ is the best possible.

It turns out that for Banach spaces the above stronger source condition is given by

$$j_q(x^\dagger) = A^* j_p(A\omega).$$

One can show the following theorem:

Theorem 4.6 (Inequality of Hein). *[29, Lemma 3.1] Let $x^\dagger$ be a minimum norm solution of $Ax = y$ such that there exists a $\omega \in X, j_q \in J_q, j_p \in J_p$ with*

$$j_q(x^\dagger) = A^* j_p(A\omega) \,.$$

Further, let x_α^δ be a minimizer of the Tikhonov functional $T_\alpha(x) := \frac{1}{p}\|Ax - y^\delta\|^p + \alpha \cdot \frac{1}{q}\|x\|^q, p, q > 1$ and $\|y - y^\delta\| \leq \delta$. Then, with $\gamma = \alpha^{1/(p-1)}$ we have

$$D_{j_q}(x^\dagger, x_\alpha^\delta) \leq D_{j_q}(x^\dagger, x^\dagger - \gamma\omega) + \alpha^{-1} \cdot D_{j_p}(-A(\gamma\omega), A(x^\dagger - \gamma\omega) - y^\delta).$$

Proof. We set $\gamma = \alpha^{1/(p-1)}$. Since

$$\frac{1}{q}\|y\|^q - \frac{1}{q}\|x\|^q = D_{j_q}(x, y) + \langle j_q(x), y - x \rangle,$$

we have

$$
\begin{aligned}
\frac{1}{q}\|x_\alpha^\delta\|^q - \frac{1}{q}\|x^\dagger - \gamma\omega\|^q &= \frac{1}{q}\|x_\alpha^\delta\|^q - \frac{1}{q}\|x^\dagger\|^q - \left(\frac{1}{q}\|x^\dagger - \gamma\omega\|^q - \frac{1}{q}\|x^\dagger\|^q\right) \\
&= D_{j_q}(x^\dagger, x_\alpha^\delta) + \langle j_q(x^\dagger), x_\alpha^\delta - x^\dagger \rangle \\
&\quad - D_{j_q}(x^\dagger, x^\dagger - \gamma\omega) - \langle j_q(x^\dagger), (x^\dagger - \gamma\omega) - x^\dagger \rangle \\
&= D_{j_q}(x^\dagger, x_\alpha^\delta) - D_{j_q}(x^\dagger, x^\dagger - \gamma\omega) \\
&\quad + \langle j_q(x^\dagger), x_\alpha^\delta + \gamma\omega - x^\dagger \rangle.
\end{aligned}
$$

Since x_α^δ is the minimizer of the Tikhonov functional, we get with the above equality that

$$
\begin{aligned}
\frac{1}{p}\|Ax_\alpha^\delta - y^\delta\|^p &+ \alpha D_{j_q}(x^\dagger, x_\alpha^\delta) \\
&= \frac{1}{p}\|Ax_\alpha^\delta - y^\delta\|^p + \alpha\frac{1}{q}\|x_\alpha^\delta\|^q - \alpha\frac{1}{q}\|x^\dagger - \gamma\omega\|^q \\
&\quad + \alpha D_{j_q}(x^\dagger, x^\dagger - \gamma\omega) - \alpha\langle j_q(x^\dagger), x_\alpha^\delta + \gamma\omega - x^\dagger \rangle \\
&\leq \frac{1}{p}\|A(x^\dagger - \gamma\omega) - y^\delta\|^p \\
&\quad + \alpha D_{j_q}(x^\dagger, x^\dagger - \gamma\omega) - \alpha\langle j_q(x^\dagger), x_\alpha^\delta + \gamma\omega - x^\dagger \rangle \,.
\end{aligned}
$$

Further, we have

$$\frac{1}{p}\|A(x^\dagger - \gamma\omega) - y^\delta\|^p = D_{j_p}(-A(\gamma\omega), A(x^\dagger - \gamma\omega) - y^\delta) + \frac{1}{p}\| - A(\gamma\omega)\|^p$$
$$+ \langle j_p(-A(\gamma\omega)), (A(x^\dagger - \gamma\omega) - y^\delta) - (-A(\gamma\omega))\rangle$$
$$= D_{j_p}(-A(\gamma\omega), A(x^\dagger - \gamma\omega) - y^\delta) + \frac{1}{p}\|A(\gamma\omega)\|^p$$
$$- \langle j_p(A(\gamma\omega)), Ax^\dagger - y^\delta\rangle$$

and

$$-\langle j_p(A(\gamma\omega)), Ax^\dagger - y^\delta\rangle + \frac{1}{p}\|A(\gamma\omega)\|^p$$
$$= -\langle j_p(A(\gamma\omega)), Ax^\dagger - Ax_\alpha^\delta + Ax_\alpha^\delta - y^\delta\rangle + \frac{1}{p}\|A(\gamma\omega)\|^p$$
$$= -\langle A^* j_p(A(\gamma\omega)), x^\dagger - x_\alpha^\delta\rangle - \langle j_p(A(\gamma\omega)), Ax_\alpha^\delta - y^\delta\rangle + \frac{1}{p}\|A(\gamma\omega)\|^p .$$

Due to Cauchy's inequality and Young's inequality, we get

$$-\langle -j_p(A(\gamma\omega)), Ax_\alpha^\delta - y^\delta\rangle + \frac{1}{p}\|A(\gamma\omega)\|^p$$
$$\leq \frac{1}{p^*}\|j_p(A(\gamma\omega))\|^{p^*} + \frac{1}{p}\|A(\gamma\omega)\|^p + \frac{1}{p}\|Ax_\alpha^\delta - y^\delta\|^p .$$

By the properties of the duality mapping, we get

$$\frac{1}{p^*}\|j_p(A(\gamma\omega))\|^{p^*} + \frac{1}{p}\|A(\gamma\omega)\|^p = \frac{1}{p^*}\|A(\gamma\omega)\|^{p^*(p-1)} + \frac{1}{p}\|A(\gamma\omega)\|^p$$
$$= \|A(\gamma\omega)\|^p$$
$$= \langle j_p(A(\gamma\omega), A(\gamma\omega)\rangle$$
$$= -\langle A^* j_p(A\gamma\omega), -\gamma\omega\rangle .$$

Hence, we have

$$-\langle j_p(A(\gamma\omega)), Ax^\dagger - y^\delta\rangle + \frac{1}{p}\|A(\gamma\omega)\|^p$$
$$\leq -\langle A^* j_p(A(\gamma\omega)), -x_\alpha^\delta - \gamma\omega + x^\dagger\rangle + \frac{1}{p}\|Ax_\alpha^\delta - y^\delta\|^p .$$

Altogether, we have

$$\frac{1}{p}\|A(x^\dagger - \gamma\omega) - y^\delta\|^p \leq D_{j_p}(-A(\gamma\omega), A(x^\dagger - \gamma\omega) - y^\delta)$$
$$-\langle A^* j_p(A(\gamma\omega)), -x_\alpha^\delta - \gamma\omega + x^\dagger\rangle + \frac{1}{p}\|Ax_\alpha^\delta - y^\delta\|^p.$$

Due to $\gamma = \alpha^{1/(p-1)}$ and by the source condition, we get

$$A^* j_p(A(\gamma\omega)) = \alpha A^* j_p(A\omega) = \alpha j_q(x^\dagger).$$

Hence, we get

$$\frac{1}{p}\|Ax_\alpha^\delta - y^\delta\|^p + \alpha D_{j_q}(x^\dagger, x_\alpha^\delta)$$
$$\leq \frac{1}{p}\|A(x^\dagger - \gamma\omega) - y^\delta\|^p + \alpha D_{j_q}(x^\dagger, x^\dagger - \gamma\omega) - \alpha\langle j_q(x^\dagger), x_\alpha^\delta + \gamma\omega - x^\dagger\rangle$$
$$\leq \alpha D_{j_q}(x^\dagger, x^\dagger - \gamma\omega) - \alpha\langle j_q(x^\dagger), x_\alpha^\delta + \gamma\omega - x^\dagger\rangle$$
$$+ D_{j_p}(-A(\gamma\omega), A(x^\dagger - \gamma\omega) - y^\delta)$$
$$- \alpha\langle j_q(x^\dagger), -x_\alpha^\delta - \gamma\omega + x^\dagger\rangle + \frac{1}{p}\|Ax_\alpha^\delta - y^\delta\|^p$$
$$= \alpha D_{j_q}(x^\dagger, x^\dagger - \gamma\omega) + D_{j_p}(-A(\gamma\omega), A(x^\dagger - \gamma\omega) - y^\delta) + \frac{1}{p}\|Ax_\alpha^\delta - y^\delta\|^p,$$

which proves the claim.   □

Remark 4.7. *If Y is a Hilbert space, then with $p = 2$ we have*

$$D_{j_p}(-A(\gamma\omega), A(x^\dagger - \gamma\omega) - y^\delta) = \frac{1}{2}\|y - y^\delta\|^2 \leq \frac{1}{2}\delta^2.$$

Hence, the inequality of Hein is a generalization of the results presented in [53, Remark and Corollary on page 1308].

Corollary 4.8. *Let $x^\dagger$ be a minimum norm solution of $Ax = y$ such that there exist a $\omega \in X, j_q \in J_q, j_p \in J_p$ such that*

$$j_q(x^\dagger) = A^* j_p(A\omega).$$

Further, let x_α^δ be a minimizer of the Tikhonov functional $\mathrm{T}_\alpha(x) := \frac{1}{p}\|Ax - y^\delta\|^p + \alpha \cdot \frac{1}{q}\|x\|^q$ and $\|y - y^\delta\| \leq \delta$ and Y is p-smooth and X is q-smooth . Then, there exists a constant $\mathcal{C} > 0$ which is independent on $x^\dagger$ and x_α^δ such that

$$D_{j_q}(x^\dagger, x_\alpha^\delta) \leq \mathcal{C}(\alpha^{q/(p-1)}\|\omega\|^q + \alpha^{-1}\delta^p).$$

Proof. Due to the inequality of Hein (cf. Theorem 4.6) and the smoothness of power type of X and Y, we have with $\gamma = \alpha^{1/(p-1)}$

$$D_{j_q}(x^\dagger, x_\alpha^\delta) \leq D_{j_q}(x^\dagger, x^\dagger - \gamma\omega) + \alpha^{-1} \cdot D_{j_p}(-A(\gamma\omega), A(x^\dagger - \gamma\omega) - y^\delta)$$
$$\leq \mathcal{C}(\|\gamma\omega\|^q + \alpha^{-1}\|Ax^\dagger - y^\delta\|^p).$$

□

4.3   Choice of the regularization parameter

The aim of this section is to show convergence rates of the form

$$D_{j_q}(x^\dagger, x_{\alpha(\delta)}^\delta) \leq \mathcal{C} \cdot \delta^\kappa$$

of the minimizers $x_{\alpha(\delta)}^{\delta}$ to the minimum norm solution $x^{\dagger}$ of $Ax = y$. To achieve this, we will consider the source conditions introduced in the last section together with an appropriate parameter choice rule for α.

First, we consider parameter choice rules of the form

$$\alpha(\delta) \sim \delta^{\nu}, \tag{4.7}$$

where $\nu > 0$. In particular, the above parameter choice rule does *not* depend on the noisy data y^{δ}. Such a parameter choice rule is called *a-priori*.

In opposite to this, for the so-called *a-posteriori* rules y^{δ} and δ have to be known, i.e. $\alpha = \alpha(\delta, y^{\delta})$. The probably most known and used a-posteriori parameter choice rule for Tikhonov functionals in Hilbert space setting is the discrepancy principle of Morozov. However, it is also well-known that the best rate which can be achieved with Morozov's discrepancy principle - in Hilbert spaces - is given by $\frac{1}{2}\|x_{\alpha(\delta, y^{\delta})}^{\delta} - x^{\dagger}\|^2 \leq C \cdot \delta$. We will show that - surprisingly - the same rate, but with respect to the Bregman distance, is also obtained if the Morozov discrepancy is applied in Banach spaces.

To obtain better convergence rates, Engl proposed in [22] a different discrepancy principle. In the last part of this section, we will show that the discrepancy principle of Engl can be generalized to the setting of Banach spaces and in the best case obtains better convergence rates than for the discrepancy principle of Morozov.

4.3.1   A-priori parameter choice

In this section, we will use a-priori parameter choice rules of the form

$$\alpha(\delta) \sim \delta^{\nu},$$

where $\nu > 0$ depends on the source condition and underlying spaces and functionals to show convergence rates of the form

$$D_{j_q}(x^{\dagger}, x_{\alpha(\delta)}^{\delta}) \leq C \cdot \delta^{\kappa}.$$

Theorem 4.9. *Let $x^{\dagger}$ be the minimum norm solution of $Ax = y$, x_{α}^{δ} the minimizer of the Tikhonov functional $T_{\alpha}(x) = \frac{1}{p}\|Ax - y^{\delta}\|^p + \alpha \cdot \frac{1}{q}\|x\|^q$ with $\|y - y^{\delta}\| \leq \delta$ and $p > 1$.*

1. If $j_q(x^{\dagger}) = A^*\omega$ for some $j_q \in J_q$ and α is given by

$$\alpha \sim \delta^{p-1},$$

   then

$$D_{j_q}(x^{\dagger}, x_{\alpha(\delta)}^{\delta}) \leq C \cdot \delta.$$

2. If $j_q(x^\dagger) = A^* j_p(A\omega)$ for some $j_q \in J_q, j_p \in J_p$, the space X is smooth of power type q, Y is smooth of power type p and α is given by

$$\alpha \sim \delta^{p(p-1)/(p+q-1)},$$

then

$$D_{j_q}(x^\dagger, x^\delta_{\alpha(\delta)}) \leq C \cdot \delta^{pq/(p+q-1)}.$$

In both cases, C neither depends on x^δ_α nor α.

Proof. First, consider the low order source condition $j_q(x^\dagger) = A^*\omega$. By Theorem 4.4, we have

$$D_{j_q}(x^\dagger, x^\delta_\alpha) \leq \tfrac{1}{p^*}\|\omega\|^{p^*} \cdot \alpha^{p^*-1} + \tfrac{1}{p}\alpha^{-1} \cdot \delta^p + \|\omega\|\delta.$$

Then, the optimal choice of $\alpha(\delta)$ is given by $\alpha \sim \delta^{p-1}$, which results in the convergence rate $D_{j_q}(x^\dagger, x^\delta_{\alpha(\delta)}) \leq C \cdot \delta$.

Next, we consider the high order source condition $j_q(x^\dagger) = A^* j_p(A\omega)$. Since the space X is smooth of power type q and Y is smooth of power type p, we get by Corollary 4.8 that

$$D_{j_q}(x^\dagger, x^\delta_\alpha) \leq C(\alpha^{q/(p-1)}\|\omega\|^q + \alpha^{-1}\delta^p).$$

Therefore, the optimal rate $\delta^{pq/(p+q-1)}$ is achieved for $\alpha \sim \delta^{p(p-1)/(p+q-1)}$. $\square$

Remark 4.10. *If additionally to the assumptions of the last theorem the space X is convex of power type, say κ-convex, then both convergence rates can be expressed with terms of norm of X.*

First, we see that

$$\alpha\tfrac{1}{q}\|x^\delta_\alpha\|^q \leq \mathrm{T}_\alpha(x^\delta_\alpha) \leq \mathrm{T}_\alpha(x^\dagger) \leq \tfrac{1}{p}\delta^p + \alpha \cdot \tfrac{1}{p}\|x^\dagger\|^p.$$

Hence, for all sufficiently small α and δ, all x^δ_α are uniformly bounded.

1. *First, assume that the low order source condition $j_q(x^\dagger) = A^*\omega$ holds. If $\kappa \leq q$, then by Corollary 2.49*

$$C\|x^\dagger - x^\delta_{\alpha(\delta)}\|^q \leq D_{j_q}(x^\dagger, x^\delta_{\alpha(\delta)}) \leq C \cdot \delta$$

and consequently

$$\|x^\dagger - x^\delta_{\alpha(\delta)}\| \leq C \cdot \delta^{1/q}.$$

If $q \leq \kappa$ then again by Corollary 2.49

$$C(\|x^\dagger\| + \|x^\delta_{\alpha(\delta)}\|)^{q-\kappa}\|x^\dagger - x^\delta_{\alpha(\delta)}\|^\kappa \leq D_{j_q}(x^\dagger, x^\delta_{\alpha(\delta)}) \leq C \cdot \delta$$

and by the boundedness of $x^\delta_{\alpha(\delta)}$ we have

$$\|x^\dagger - x^\delta_{\alpha(\delta)}\| \leq C \cdot \delta^{1/\kappa}$$

for all sufficiently small δ.

2. *Assume on the other hand that the high order source condition $j_q(x^\dagger) = A^* j_p(A\omega)$ holds, where X is q-smooth. Then, by Theorem 2.41 and 2.42 we know that $q \leq \kappa$. Hence, once again by Corollary 2.49, we get*

$$\|x^\dagger - x^\delta_{\alpha(\delta)}\| \leq \mathcal{C} \cdot \delta^{pq/[(p+q-1)\kappa]}$$

for sufficiently small δ.

4.3.2   Morozov's discrepancy principle

In the last section, we have shown how the parameter choice rule $\alpha(\delta)$ can be chosen without using the information about y^δ, i.e. we have discussed *a-priori* rules. In this section, we will present one of the most known and used parameter choice rules which employ the information about y^δ, namely, the discrepancy principle of Morozov.

The parameter α is chosen appropriately if the minimizer x^δ_α resembles the minimum norm solution x^δ, i.e. $x^\delta_\alpha \approx x^\dagger$. We know that $\|Ax^\dagger - y^\delta\| = \|y - y^\delta\| \approx \delta$. Assuming that $x^\delta_\alpha \approx x^\dagger$ we may conclude that

$$\|Ax^\delta_\alpha - y^\delta\| \approx \|Ax^\dagger - y^\delta\| = \|y - y^\delta\| \approx \delta$$

holds for a good choice of α.

Hence, the idea of the discrepancy principle is to chose $\alpha(\delta, y^\delta)$ such that

$$\|Ax^\delta_{\alpha(\delta,y^\delta)} - y^\delta\| = \tau\delta,$$

where the parameter τ has to be chosen appropriately to ensure that such $\alpha(\delta, y^\delta)$ exists. However, we do not want to go into detail of the proof when such parameter exists. For the remaining part of the section we will assume that this is the case. However, we remark that the proof of existence can be carried out along the lines of the proofs of [37] or [6]. The following result can be proven:

Theorem 4.11. *Let $x^\dagger$ be a minimum norm solution of $Ax = y$ such that $j_q(x^\dagger) = A^*\omega$ for some $j_q \in J_q$ and $\omega \in Y^*$, x^δ_α be a minimizer of the Tikhonov functional $T_\alpha(x) := \frac{1}{p}\|Ax - y^\delta\|^p + \alpha \cdot \frac{1}{q}\|x\|^q$, where $p > 1$. Further, let $\tau > 1$ be chosen such that for y^δ with $\|y - y^\delta\| \leq \delta$ there exists an $\alpha(\delta, y^\delta)$ such that*

$$\|Ax^\delta_{\alpha(\delta,y^\delta)} - y^\delta\| = \tau\delta.$$

Then,

$$D_{j_q}(x^\dagger, x^\delta_{\alpha(\delta,y^\delta)}) \leq (1+\tau)\|\omega\| \cdot \delta.$$

Proof. By Theorem 4.4 we have

$$D_{j_q}(x^\dagger, x^\delta_\alpha) \leq \alpha^{-1} \cdot \frac{1}{p}(\delta^p - \|Ax^\delta_\alpha - y^\delta\|^p) + \|\omega\|(\|Ax^\delta_\alpha - y^\delta\| + \delta).$$

Since $\|Ax^{\delta}_{\alpha(\delta,y^{\delta})} - y^{\delta}\| = \tau\delta$ with $\tau > 1$, we have

$$\delta^p - \|Ax^{\delta}_{\alpha(\alpha,y^{\delta})} - y^{\delta}\|^p = \delta^p - (\tau\delta)^p \leq 0.$$

Which proves the claim, since $\|Ax^{\delta}_{\alpha(\delta,y^{\delta})} - y^{\delta}\| + \delta = \tau\delta + \delta.$ □

If X and Y are Hilbert spaces and $q = 2$, then the convergence rate obtained above is the best possible for Morozov's discrepancy principle, cf. [23, Section 4.3]. This fact is known as *early saturation* of Morozov's discrepancy principle. Following the common knowledge that with the transition to Banach spaces the convergence can only get worse, the author assumes that this is also the best possible rate in Banach space setting. This premature saturation is the main drawback of the discrepancy principle of Morozov.

4.3.3   Engl's discrepancy principle

We still consider the Tikhonov functional

$$T_\alpha(x) := \tfrac{1}{p}\|Ax - y^{\delta}\|^p + \alpha\tfrac{1}{q}\|x\|^q \qquad p, q > 1.$$

In [22] it was shown that the choice of $\alpha(\delta, y^{\delta})$ via

$$\|A^* j^Y_p (Ax^{\delta}_{\alpha(\delta,y^{\delta})} - y^{\delta})\|^{q^*}_{X^*} = \delta^r \alpha^{-s} \qquad r, s > 0. \tag{4.8}$$

leads to an optimal convergence rate in the Hilbert space setting if the parameters r and s are chosen appropriately. Since then Engl's modified discrepancy principle was extended to non-linear operator equations in [38]. In this section, we will show that Engl's approach can also be extended to linear operators mapping between Banach spaces.

We will show the claims as an series of theorems. First, in Theorem 4.13, we will show that Engl's discrepancy principle is well-defined in Banach spaces. In Theorem 4.15, we will show that it is a regularization and finally in Theorem 4.16 we will show that the same convergence rates are achieved as for a-priori parameter choice if one of the source conditions

$$j_q(x^{\dagger}) = A^*\omega \qquad \omega \in Y^*$$

or

$$j_q(x^{\dagger}) = A^* j_p(A\omega) \qquad \omega \in X$$

holds.

Remark 4.12. *We remark that all results of this section do also hold if $\alpha(\delta, y^{\delta})$ is chosen in a more general way. Namely, if $\alpha(\delta, y^{\delta})$ is chosen such that*

$$\|A^* j^Y_p (Ax^{\delta}_{\alpha(\delta,y^{\delta})} - y^{\delta})\|^{q^*}_{X^*} \sim \delta^r \alpha^{-s},$$

i.e.

$$c \cdot \delta^r \alpha^{-s} \leq \|A^* j_p^Y (A x_{\alpha(\delta, y^\delta)}^\delta - y^\delta)\|_{X^*}^{q^*} \leq C \cdot \delta^r \alpha^{-s},$$

where the constants $c, C > 0$ *are independent of* δ *and* y^δ.

However, for the sake of clarity we will prove the claims for c and C being both set to 1.

As announced we begin with the well-definition of (4.8).

Theorem 4.13. *Let* X *be convex of power type,* Y *be smooth of power type. Then, for every* $r, s > 0, p, q > 1, \delta > 0, y^\delta \in Y$ *with* $A^* j_p(y^\delta) \neq 0$ *there exists a* $\alpha > 0$ *such that*

$$\|A^* j_p^Y (A x_\alpha^\delta - y^\delta)\|_{X^*}^{q^*} = \delta^r \alpha^{-s},$$

i.e. the parameter choice (4.8) *is well-defined.*

Proof. We will show the claim in three steps

1. First, we show that $\alpha^s \|A^* j_p (A x_\alpha^\delta - y^\delta)\|^{q^*} \to 0$ as $\alpha \to 0$.

2. Then, we will show that $\alpha^s \|A^* j_p (A x_\alpha^\delta - y^\delta)\|^{q^*} \to \infty$ as $\alpha \to \infty$.

3. Finally, we will show that the function $\alpha \mapsto \alpha^s \|A^* j_p (A x_\alpha^\delta - y^\delta)\|^{q^*}$ is continuous.

First, we notice that

$$A^* J_p(A x_\alpha^\delta - y^\delta) = \partial(\tfrac{1}{p} \|A \cdot - y^\delta\|^p)(x_\alpha^\delta) \; ;$$

and due to smoothness of power type of Y the duality mapping J_p is single valued.

The optimality condition for T_α is given by $0 \in A^* J_p(A x_\alpha^\delta - y^\delta) + \alpha \cdot J_q(x_\alpha^\delta)$. Therefore, there exists a $j_q \in J_q$ such that

$$0 = A^* j_p(A x_\alpha^\delta - y^\delta) + \alpha \cdot j_q(x_\alpha^\delta).$$

Hence, we have

$$\|A^* j_p(A x_\alpha^\delta - y^\delta)\|^{q^*} = \alpha^{q^*} \|x_\alpha^\delta\|^q \leq \alpha^{q^*-1} q \, T_\alpha(x_\alpha^\delta) \leq \alpha^{q^*-1} q \, T_\alpha(x^\dagger).$$

Therefore, for bounded α the term $\|A^* j_p(A x_\alpha^\delta - y^\delta)\|^{q^*}$ is bounded. Hence, $\alpha^s \|A^* j_p(A x_\alpha^\delta - y^\delta)\|^{q^*} \longrightarrow 0$ as $\alpha \longrightarrow 0$.

Further, we have

$$
\begin{aligned}
\|x_\alpha^\delta\|^{q-1} &\leq \alpha^{-1} \|A^*\| \|A x_\alpha^\delta - y^\delta\|^{p-1} \\
&\leq \mathcal{C} \cdot \alpha^{-1} T_\alpha(x_\alpha^\delta)^{\frac{p-1}{p}} \\
&\leq \mathcal{C} \cdot \alpha^{-1} T_\alpha(x^\dagger)^{\frac{p-1}{p}} \\
&\leq \mathcal{C} \cdot \alpha^{-1}(\mathcal{C} + \alpha \cdot \mathcal{C})^{\frac{p-1}{p}} \\
&\leq \mathcal{C} \cdot (\alpha^{-1} + \alpha^{-1/p})
\end{aligned}
$$

Therefore, we have $x_\alpha^\delta \longrightarrow 0$ as $\alpha \longrightarrow \infty$.

The space Y is assumed to be smooth of power type, say ρ-smooth. Then, by the Xu-Roach inequalities (cf. Theorem 2.33) we have

$$\begin{aligned}
\|A^* j_p(Ax_\alpha^\delta - y^\delta) - A^* j_p(y^\delta)\| &\leq C \cdot \|j_p(Ax_\alpha^\delta - y^\delta) - j_p(y^\delta)\| \\
&\leq C \cdot (\max\{\|Ax_\alpha^\delta - y^\delta\|, \|y^\delta\|\})^{p-\rho} \|x_\alpha^\delta\|^{\rho-1}.
\end{aligned}$$

If $p \geq \rho$, then, since $x_\alpha^\delta \longrightarrow 0$ as $\alpha \longrightarrow \infty$, there exists a constant $C > 0$ such that for all α big enough

$$(\max\{\|Ax_\alpha^\delta - y^\delta\|, \|y^\delta\|\})^{p-\rho} \leq (\|A\|\|x_\alpha^\delta\| + \|y^\delta\|)^{p-\rho} \leq C.$$

If on the other hand $p < \rho$, then, as a consequence of $A^* j_p(y^\delta) \neq 0$, we know that $y^\delta \neq 0$. Therefore, there exists a constant $C > 0$ such that

$$(\max\{\|Ax_\alpha^\delta - y^\delta\|, \|y^\delta\|\})^{p-\rho} \leq 1/\|y^\delta\|^{\rho-p} \leq C.$$

Hence, for $\alpha \to \infty$ we always have

$$\|A^* j_p(Ax_\alpha^\delta - y^\delta)\|^{q^*} \longrightarrow \|A^* j_p(y^\delta)\|^{q^*} > 0$$

and consequently $\alpha^s \|A^* j_p(Ax_\alpha^\delta - y^\delta)\|^{q^*} \longrightarrow \infty$ as $\alpha \longrightarrow \infty$.

Finally, we show that the mapping $\alpha \mapsto \alpha^s \|A^* j_p(Ax_\alpha^\delta - y^\delta)\|^{q^*}$ is continuous. Since $\|A^* j_p(Ax_\alpha^\delta - y^\delta)\| = \alpha \|x_\alpha^\delta\|^{q-1}$ this is true if $\alpha \mapsto \|x_\alpha^\delta\|$ is continuous.

Assume that $\alpha > 0$ is fixed. Since X is convex of power type and since J_p is the subgradient of $\frac{1}{p} \| \cdot \|^p$, we have for every $\beta > 0$

$$\begin{aligned}
\frac{1}{p}\|Ax_\alpha^\delta - y^\delta\|^p &\geq \frac{1}{p}\|Ax_\beta^\delta - y^\delta\|^p + \langle j_p(Ax_\beta^\delta - y^\delta), Ax_\alpha^\delta - Ax_\beta^\delta\rangle \\
\alpha\frac{1}{q}\|x_\alpha^\delta\|^q &\geq \alpha\frac{1}{q}\|x_\beta^\delta\|^q + \alpha\langle j_q(x_\beta^\delta), x_\alpha^\delta - x_\beta^\delta\rangle + \alpha\sigma_q(x_\beta^\delta, x_\beta^\delta - x_\alpha^\delta),
\end{aligned}$$

where the function $\sigma_q > 0$ is defined as in Theorem 2.32. Summing up the inequalities and due to the optimality condition we get

$$T_\alpha(x_\alpha^\delta) \geq T_\alpha(x_\beta^\delta) + (\alpha - \beta)\langle j_q(x_\beta^\delta), x_\alpha^\delta - x_\beta^\delta\rangle + \alpha\sigma_q(x_\beta^\delta, x_\beta^\delta - x_\alpha^\delta).$$

Since x_α^δ minimizes T_α we have $0 \geq T_\alpha(x_\alpha^\delta) - T_\alpha(x_\beta^\delta)$. Hence, we have

$$0 \leq \alpha\sigma_p(x_\beta^\delta, x_\beta^\delta - x_\alpha^\delta) \leq (\alpha - \beta)\langle j_q(x_\beta^\delta), x_\alpha^\delta - x_\beta^\delta\rangle \leq (\alpha - \beta)\|x_\beta^\delta\|^{q-1}\|x_\alpha^\delta - x_\beta^\delta\|.$$

Further, for all sufficiently small $\delta > 0$, and all β sufficiently close to α we have $\|x_\beta^\delta\| \leq C$, cf. Remark 4.10. Therefore,

$$\alpha\sigma_q(x_\beta^\delta, x_\beta^\delta - x_\alpha^\delta) \leq |\beta - \alpha| \cdot \|x_\beta^\delta\|^{q-1}\|x_\alpha^\delta - x_\beta^\delta\| \leq C|\beta - \alpha| \cdot \|x_\alpha^\delta - x_\beta^\delta\|.$$

If $q - \kappa \leq 0$, then by Corollary 2.49 of the Xu-Roach inequalities for convexity of power type and the uniform boundedness of the minimizers of the Tikhonov function for all sufficiently small $\alpha, \beta, \delta > 0$, we have

$$\sigma_q(x_\beta^\delta, x_\beta^\delta - x_\alpha^\delta) \geq C(\|x_\alpha^\delta\| + \|x_\beta^\delta\|)^{p-\kappa}\|x_\alpha^\delta - x_\beta^\delta\|^\kappa \geq C\|x_\alpha^\delta - x_\beta^\delta\|^\kappa$$

and we get

$$\alpha \cdot C\|x_\alpha^\delta - x_\beta^\delta\|^{\kappa-1} \leq |\beta - \alpha|.$$

Hence, $\|x_\beta^\delta\| \longrightarrow \|x_\alpha^\delta\|$ as $\beta \longrightarrow \alpha > 0$.

If on the other hand $q - \kappa > 0$, then again by Corollary 2.49

$$\sigma_q(x_\beta^\delta, x_\beta^\delta - x_\alpha^\delta) \geq C\|x_\alpha^\delta - x_\beta^\delta\|^q.$$

Hence,

$$\alpha \cdot C\|x_\alpha^\delta - x_\beta^\delta\|^{q-1} \leq |\beta - \alpha|.$$

And therefore, again $\|x_\beta^\delta\| \longrightarrow \|x_\alpha^\delta\|$ as $\beta \longrightarrow \alpha > 0$. This completes the proof. $\square$

Remark 4.14. *If Y is smooth of power type and $A^* J_p(y) \neq 0$, then for all sufficiently small δ we also have $A^* J_p(y^\delta) \neq 0$. Since J_p is Hölder continuous on bounded subsets of spaces smooth of power type, cf. Corollary 2.35. Hence, if $A^* J_p(y) \neq 0$, then the assumption of the last theorem is fulfilled for sufficiently small $\delta > 0$. In the next theorem, we will see that this condition is also sufficient to show the regularization property.*

Theorem 4.15. *Let X be convex of power type, Y be smooth of power type, $x^\dagger$ be the minimum norm solution of $Ax = y$ with $y \in \mathcal{R}(A)$, where $A^* j_p(y) \neq 0$. Further, let $r, s > 0, p, q > 1$ be such that*

$$(s + q^* - 1)p - r \geq 0,$$

then

$$x_{\alpha(\delta, y^\delta)}^\delta \longrightarrow x^\dagger \quad as \quad \delta \longrightarrow 0.$$

i.e. the parameter choice is regularizing.

Proof. Since Y is smooth of power type the duality mapping J_p is single valued and continuous on bounded sets, cf. Corollary 2.35. Therefore, by Theorem 4.13 the parameter choice (4.8) is well-defined for all sufficiently small δ.

We will show the claim in two steps:

1. First, we will show that $\alpha(\delta, y^\delta) \to 0$ as $\delta \to 0$.

2. Further, we will show that $\delta^p / \alpha(\delta, y^\delta) \to 0$ as $\delta \to 0$.

Then, the claim is a consequence of Theorem 4.2.

Due to the optimality condition we have

$$\alpha(\delta, y^\delta)^{q^*} \|x^\delta_{\alpha(\delta,y^\delta)}\|^q = \|A^* j_p(Ax^\delta_{\alpha(\delta,y^\delta)} - y^\delta)\|^{q^*} = \delta^r \alpha(\delta, y^\delta)^{-s} \qquad (4.9)$$

Assume there is a sequence $\delta_n \to 0$ as $n \to 0$ such that for $\alpha_n := \alpha(\delta_n, y^{\delta_n})$ we have $\limsup_{n\to\infty} \alpha_n \geq c > 0$. Without restriction, we can assume that $\lim_{n\to\infty} \alpha_n = c$ and that $\alpha_n > c/2$ for all n. Due to (4.9) we have

$$\|x^{\delta_n}_{\alpha_n}\|^q = \delta_n^r \alpha_n^{-(s+q^*)} \leq \delta_n^r (c/2)^{-(s+q^*)} \longrightarrow 0.$$

Since j_p is continuous on bounded sets we get that

$$\begin{aligned}
0 &= \lim_{n\to\infty} \delta_n^r (c/2)^{-s} \geq \lim_{n\to\infty} \delta_n^r \alpha_n^{-s} \\
&= \lim_{n\to\infty} \|A^* j_p(Ax^{\delta_n}_{\alpha_n} - y^{\delta_n})\|^{q^*} = \|A^* j_p(y)\|^{q^*} > 0.
\end{aligned}$$

Hence, we get $\alpha(\delta, y^\delta) \longrightarrow 0$ as $\delta \longrightarrow 0$.

Further, we have

$$\begin{aligned}
\delta^r &= \alpha(\delta, y^\delta)^{s+q^*} \|x^\delta_{\alpha(\delta,y^\delta)}\|^q \leq \alpha(\delta, y^\delta)^{s+q^*-1} q\, T_{\alpha(\delta,y^\delta)}(x^\delta_{\alpha(\delta,y^\delta)}) \\
&\leq \alpha(\delta, y^\delta)^{s+q^*-1} q\, T_{\alpha(\delta,y^\delta)}(x^\dagger)
\end{aligned}$$

and

$$T_{\alpha(\delta,y^\delta)}(x^\dagger) \leq \tfrac{1}{p}\delta^p + \alpha(\delta, y^\delta) \cdot \tfrac{1}{q}\|x^\dagger\|^q \longrightarrow 0 \quad \text{as} \quad \delta \longrightarrow 0$$

and therefore

$$(\delta^p/\alpha(\delta, y^\delta))^r \leq (q\, T_{\alpha(\delta,y^\delta)}(x^\dagger))^p \cdot \alpha(\delta, y^\delta)^{(s+q^*-1)p-r} \longrightarrow 0 \quad \text{as} \quad \delta \to 0,$$

since by assumption $(s+q^*-1)p - r \geq 0$. Hence, $\delta^p/\alpha(\delta, y^\delta) \longrightarrow 0$ as $\delta \longrightarrow 0$. $\square$

Theorem 4.16. *Let the parameter choice (4.8) be well-defined and regularizing (i.e. $\alpha(\delta, y^\delta)$ exists and $x^\delta_{\alpha(\delta,y^\delta)} \to x^\dagger$ for $\delta \to 0$).*

1. *If $j_q(x^\dagger) = A^* \omega$ for some $j_q \in J_q$ and $\omega \in Y^*$ and $r, s > 0$ are chosen such that*

$$\frac{r}{s+q^*} = p - 1.$$

Then, for sufficiently small δ we have

$$D_{j_q}(x^\dagger, x^\delta_{\alpha(\delta,y^\delta)}) \leq C \cdot \delta.$$

2. *If Y is p-smooth, X is q-smooth where p, q are the powers in $\mathrm{T}_\alpha(x) = \frac{1}{p}\|Ax - y^\delta\|^p + \alpha\frac{1}{q}\|x\|^q$ and further the source condition $j_q(x^\dagger) = A^* j_p(A\omega)$ holds for some $j_q \in J_q, j_p \in J_p$ and $\omega \in X$ and $r, s > 0$ are chosen such that*

$$\frac{r}{s + q^*} = \frac{p}{p + q - 1} \cdot (p - 1),$$

then for sufficiently small δ we have

$$D_{j_q}(x^\dagger, x^\delta_{\alpha(\delta, y^\delta)}) \leq C \cdot \delta^{\frac{pq}{p+q-1}}.$$

Proof. First, we consider the case $x^\dagger \neq 0$. We remark that due to the optimality condition of T_α we have

$$
\begin{aligned}
\delta^r \alpha(\delta, y^\delta)^{-(s+q^*)} &= \alpha(\delta, y^\delta)^{-q^*} \|A^* j_p(Ax^\delta_{\alpha(\delta, y^\delta)} - y^\delta)\|^{q^*} \\
&= \alpha(\delta, y^\delta)^{-q^*} \alpha(\delta, y^\delta)^{q^*} \|x^\delta_{\alpha(\delta, y^\delta)}\|^q = \|x^\delta_{\alpha(\delta, y^\delta)}\|^q.
\end{aligned}
$$

Therefore, due to the regularization property we have

$$\lim_{\delta \to 0} \delta^r \cdot \alpha(\delta, y^\delta)^{-(s+q^*)} = \lim_{\delta \to 0} \|x^\delta_{\alpha(\delta, y^\delta)}\|^q = \|x^\dagger\|^q > 0.$$

Hence, for all sufficiently small $\delta > 0$ we have

$$0 < c \leq \delta^r \cdot \alpha(\delta, y^\delta)^{-(s+q^*)} \leq C < \infty. \tag{4.10}$$

We now show the first claim. By Theorem 4.4 and (4.10) we have

$$
\begin{aligned}
D_{j_q}(x^\dagger, x^\delta_{\alpha(\delta, y^\delta)}) &\leq C\left(\delta^p \alpha(\delta, y^\delta)^{-1} + \alpha(\delta, y^\delta)^{\frac{1}{p-1}} + \delta\right) \\
&\leq C\left(\delta^{p - \frac{r}{s+q^*}} + \delta^{\frac{r}{s+q^*} \cdot \frac{1}{p-1}} + \delta\right).
\end{aligned}
$$

The claim follows, since $\frac{r}{s+q^*} = p - 1$.

Next, we show the second claim. With the Corollary 4.8 of the inequality of Hein and (4.10) we have

$$
\begin{aligned}
D_{j_q}(x^\dagger, x^\delta_{\alpha(\delta, y^\delta)}) &\leq C\left(\alpha(\delta, y^\delta)^{\frac{q}{p-1}} + \alpha(\delta, y^\delta)^{-1} \cdot \delta^p\right) \\
&\leq C\left(\delta^{\frac{r}{s+q^*} \cdot \frac{q}{p-1}} + \delta^{p - \frac{r}{s+q^*}}\right).
\end{aligned}
$$

The claim follows since this time $\frac{r}{s+q^*} = \frac{p}{p+q-1} \cdot (p-1)$.

Finally, we consider the case $x^\dagger = 0$. We have

$$\lim_{\delta \to 0} \delta^r \cdot \alpha(\delta, y^\delta)^{-(s+q^*)} = \lim_{\delta \to 0} \|x^\delta_{\alpha(\delta, y^\delta)}\|^{q^*} = \|x^\dagger\|^{q^*} = 0.$$

Hence, for all sufficiently small $\delta > 0$ we have

$$0 \leq \delta^r \cdot \alpha(\delta, y^\delta)^{-(s+q^*)} \leq C < \infty.$$

As in the proof of Theorem 4.4, we may conclude that

$$\alpha D_{j_q}(x^\dagger, x_\alpha^\delta) \leq \tfrac{1}{p}\delta^p - \langle j_q(x^\dagger), x_\alpha^\delta - x^\dagger \rangle.$$

Since $x^\dagger = 0$, we get $j_q(x^\dagger) = 0$ and therefore

$$D_{j_q}(x^\dagger, x_{\alpha(\delta,y^\delta)}^\delta) \leq \tfrac{1}{p}\delta^p \cdot \alpha(\delta, y^\delta)^{-1} \leq C\delta^{p-r/(s+q^*)}.$$

Hence, the claims hold in this case too. $\qquad\qquad\qquad\qquad\qquad\qquad\qquad\square$

4.4 Minimization of the Tikhonov functional

In this section, we consider the problem

$$\min_{x \in X} T_\alpha(x) \qquad\qquad\qquad (4.11)$$

of minimizing the Tikhonov functional $T_\alpha : X \to \mathbb{R}$ given by

$$T_\alpha(x) := \tfrac{1}{p}\|Ax - y^\delta\|_Y^p + \alpha\tfrac{1}{q}\|x\|_X^q \qquad p, q > 1,$$

where a continuous linear operator $A : X \to Y$ mapping between two Banach spaces X and Y.

If X and Y are Hilbert spaces, then T_α is Gâteaux differentiable and one can show that the well-known steepest descent iteration given by

$$x_{n+1} = x_n - \mu_n \nabla T_\alpha(x_n)$$

converges to the unique minimizer of problem (4.11) if the step size μ_n is chosen properly. The above iteration cannot be trivially generalized to the setting of Banach spaces, since then the iterates x_n are elements of the primal space X but the gradient $\nabla T_\alpha(x_n)$ is an element of the dual space X^*. However, we remember that with the duality mappings J_q^X and $J_{q^*}^{X^*}$ we have two function at hand, which we can use to transport elements from the primal space into the dual space and from the dual space back into the primal space.

Therefore, we could use the mapping $J_{q^*}^{X^*}$ to transport the gradient $\nabla T_\alpha(x_n)$ back into the primal space. Then, the resulting iteration is given by

$$x_{n+1} = x_n - \mu_n J_{q^*}^{X^*}(\nabla T_\alpha(x_n)).$$

We will call the above iteration *steepest descent (with update) in the primal (space)*.

At least formally, we could also start with a dual iterate, which we denote by x_n^*, then update the gradient in dual space and finally transport the resulting dual iterate x_{n+1}^* back into the primal space by $J_{q^*}^{X^*}$. The resulting iteration is given by

$$
\begin{aligned}
x_{n+1}^* &= x_n^* - \mu_n \nabla \mathrm{T}_\alpha(x_n) \\
x_{n+1} &= J_q^{X^*}(x_{n+1}^*) \ .
\end{aligned}
$$

For obvious reasons we will denote the above iteration as *steepest descent (with update) in the dual (space)*.

For both iteration types the aim of this section is to introduce sensible step sizes μ_n, such that the iterates converge strongly to the minimizer of the functional T_α.

In the first part of the section, we will consider the update in the primal space. We will show that if the space Y is p-smooth and X is q-smooth and c-convex , where p and q are the indices in the Tikhonov functional T_α, then - for appropriately chosen μ_n - the convergence rate

$$
\|x_n - x_\alpha^\delta\| \leq \mathcal{C} \cdot n^{-(\min\{p,q\}-1)/(c-\min\{p,q\})},
$$

holds as long as either $p < c$ or $q < c$. Further, for the case that $p = q = c = 2$ we will show that the convergence rate

$$
\|x_n - x_\alpha^\delta\| \leq \mathcal{C} \cdot \exp(-n/\mathcal{C})
$$

holds.

In the second part of the section, we will consider for the update in the dual space. We will show that if the space X is q-convex and Y arbitrarily chosen, where again the index q is the same as in the definition of the Tikhonov functional, then - for appropriately chosen μ_n - the convergence rate

$$
\|x_n - x_\alpha^\delta\| \leq \mathcal{C} \cdot n^{-\frac{1}{q(q-1)}}
$$

holds. Further, if additionally the space Y is assumed to be p-convex, then the rate

$$
\|x_n - x_\alpha^\delta\| \leq \mathcal{C} \cdot n^{-\frac{(M-1)}{[(M-1)(q-1)-1]q}},
$$

where

$$
M := \max\{p, q\}
$$

holds as long as $M > 2$. For $M = 2$ the rate improves again to

$$
\|x_n - x_\alpha^\delta\| \leq \mathcal{C} \cdot \exp(-n/\mathcal{C}).
$$

4.4.1 Update in the primal space

In this section, we consider minimizing the functional

$$T_\alpha(x) = \frac{1}{p}\|Ax - y^\delta\|_Y^p + \alpha\frac{1}{q}\|x\|_X^q,$$

where the powers p, q are connected to the properties of X and Y. In particular, throughout this section we assume that

$$Y \text{ is } p\text{-smooth} \qquad \text{and} \qquad X \text{ is } q\text{-smooth.}$$

Further, we assume that

$$X \text{ is } c\text{-convex.}$$

In [7] the authors have shown that the iteration defined via

$$x_{n+1} := x_n - \mu_n J_{q^*}^{X^*}\psi_n, \qquad \psi_n = \nabla T_\alpha(x_n) = A^* J_p^Y(Ax_n - y^\delta) + \alpha J_q^X(x_n),$$

where μ_n is chosen such that

$$\mu_n := \operatorname{argmin}_\mu T_\alpha(x_n - \mu J_{q^*}^{X^*}(\psi_n))$$

converges strongly to the unique minimizer x_α^δ of the Tikhonov functional T_α. In this section, we will show which convergence rates are achieved by this choice of the step size.

Remark 4.17. *We will see that the convergence results obtained in this section hold also for the step size*

$$\mu_n := \operatorname{argmin}_\mu\{-\mu\|\psi_n\|^{q^*} + |\mu|^p\frac{G_Y}{p}\|A J_{q^*}^{X^*}(\psi_n)\|^p + \alpha|\mu|^q\frac{G_X}{q}\|\psi_n\|^{q^*}\},$$

where G_X and G_Y are the constants in the definition of smoothness of power type for X and Y. We notice that, to compute the above modified step size, we only have to solve a one-dimensional minimization problem. However, we also have to know the values of the constants G_X and G_Y, which are not needed in the original step size.

Theorem 4.18. *Let X be q-smooth and c-convex Banach space, Y p-smooth Banach space and*

$$s := \min\{p, q\} < c.$$

Then,

$$\|x_n - x_\alpha^\delta\| \le C \cdot n^{-(s-1)/(c-s)}.$$

Proof. Since X is q-smooth and is Y is p-smooth there exist constants $G_X, G_Y > 0$ such that

$$\frac{1}{p}\|x - y\|_Y^p \le \frac{1}{p}\|x\|_Y^p - \langle J_p^Y(x), y\rangle + \frac{G_Y}{p}\|y\|_Y^p \qquad \forall x, y \in Y \quad \text{and}$$
$$\frac{1}{q}\|x - y\|_X^q \le \frac{1}{q}\|x\|_X^q - \langle J_q^X(x), y\rangle + \frac{G_X}{q}\|y\|_X^q \qquad \forall x, y \in X.$$

Hence, for every μ we get that

$$\frac{1}{p}\|A(x_n - \mu J_{q^*}^{X^*}(\psi_n)) - y^\delta\|^p$$
$$\leq \frac{1}{p}\|Ax_n - y^\delta\|^p - \mu\langle J_p^Y(Ax_n - y^\delta), AJ_{q^*}^{X^*}(\psi_n)\rangle$$
$$+ |\mu|^p \frac{G_Y}{p}\|AJ_{q^*}^{X^*}(\psi_n)\|^p$$
$$= \frac{1}{p}\|Ax_n - y^\delta\|^p - \mu\langle A^* J_p^Y(Ax_n - y^\delta), J_{q^*}^{X^*}(\psi_n)\rangle$$
$$+ |\mu|^p \frac{G_Y}{p}\|AJ_{q^*}^{X^*}(\psi_n)\|^p$$

and

$$\frac{1}{q}\|x_n - \mu J_{q^*}^{X^*}(\psi_n)\|^q$$
$$\leq \frac{1}{q}\|x_n\|^q - \mu\langle J_q^X(x_n), J_{q^*}^{X^*}(\psi_n)\rangle + |\mu|^q \frac{G_X}{q}\|J_{q^*}^{X^*}(\psi_n)\|_X^q$$
$$= \frac{1}{q}\|x_n\|^q - \mu\langle J_q^X(x_n), J_{q^*}^{X^*}(\psi_n)\rangle + |\mu|^q \frac{G_X}{q}\|\psi_n\|_X^{q^*}.$$

With the equalities above, we get that

$$\mathrm{T}_\alpha(x_n - \mu\psi_n) \leq \mathrm{T}_\alpha(x_n) - \mu\|\psi_n\|^{q^*}$$
$$+ |\mu|^p \frac{G_Y}{p}\|AJ_{q^*}^{X^*}(\psi_n)\|^p + \alpha|\mu|^q \frac{G_X}{q}\|\psi_n\|^{q^*}, \tag{4.12}$$

since $\langle A^* J_p^Y(Ax_n - y^\delta) + \alpha J_q^X(x_n), J_{q^*}^{X^*}(\psi_n)\rangle = \langle\psi_n, J_{q^*}^{X^*}(\psi_n)\rangle = \|\psi_n\|^{q^*}$.
The iterates x_n are uniformly bounded, since

$$\alpha\frac{1}{q}\|x_n\|^q \leq \mathrm{T}_\alpha(x_n) \leq \mathrm{T}_\alpha(x_{n-1}) \leq \ldots \leq \mathrm{T}_\alpha(x_0).$$

Since X and Y are smooth of power type q resp. p, the mappings $x \mapsto j_q(x)$ and $x \mapsto A^* j_p(Ax - y^\delta)$ are $(q-1)$ resp. $(p-1)$ Hölder continuous, cf. Theorem 2.33. Therefore, we have

$$\|\psi_n\| \leq \mathcal{C} \cdot (\|x_n - x_\alpha^\delta\|^{p-1} + \|x_n - x_\alpha^\delta\|^{q-1}).$$

Consequently, the gradients ψ_n are uniformly bounded. With $s = \min\{p, q\}$ we have

$$\min(p(q^* - 1), q^*) = \min(p(q^* - 1), q(q^* - 1)) = s(q^* - 1).$$

Since $\|\psi_n\|$ are uniformly bounded we can estimate

$$\max\{\frac{G_Y}{p}\|AJ_{q^*}^{X^*}(\psi_n)\|^p, \alpha\frac{G_X}{q}\|\psi_n\|^{q^*}\} \leq \mathcal{C} \cdot \max\{\|A\|\|\psi_n\|^{(q^*-1)p}, \|\psi_n\|^{q^*}\}$$
$$\leq \mathcal{C} \cdot \max(\{\|\psi_n\|^{(q^*-1)p}, \|\psi_n\|^{q^*}\}$$
$$\leq \mathcal{C} \cdot \|\psi_n\|^{s(q^*-1)}.$$

Moreover, we get

$$
\begin{aligned}
\mathrm{T}_\alpha(x_{n+1}) &= \min_\mu \mathrm{T}_\alpha(x_n - \mu\psi_n) \\
&\leq \min_\mu \{\mathrm{T}_\alpha(x_n) - \mu\|\psi_n\|^{q^*} + (|\mu|^p + |\mu|^q) \cdot \|\psi_n\|^{s(q^*-1)} \cdot \mathcal{C}\} \\
&\leq \min_{0<\mu<1} \{\mathrm{T}_\alpha(x_n) - \mu\|\psi_n\|^{q^*} + (\mu^p + \mu^q) \cdot \|\psi_n\|^{s(q^*-1)} \cdot \mathcal{C}\}
\end{aligned}
$$

For $0 < \mu < 1$ we have $\mu^p \leq \mu^s$ and $\mu^q \leq \mu^s$, since $s \leq p$ and $s \leq q$. Altogether we get that

$$
\mathrm{T}_\alpha(x_{n+1}) \leq \min_{0<\mu<1} \{\mathrm{T}_\alpha(x_n) - \mu\|\psi_n\|^{q^*} + \mu^s \cdot \|\psi_n\|^{s(q^*-1)} \cdot C_0\} \tag{4.13}
$$

for some $C_0 > 0$. The minimization problem on the right-hand side of the last inequality is solved by

$$
m_n := \min\left\{1, 1/(sC_0)^{1/(s-1)} \cdot \|\psi_n\|^{s^*-q^*}\right\}.
$$

If $m_n = 1/(sC_0)^{1/(s-1)} \cdot \|\psi_n\|^{s^*-q^*}$, then

$$
-m_n \cdot \|\psi_n\|^{q^*} + m_n^s \cdot \|\psi_n\|^{s(q^*-1)} \cdot C_0 \leq -\mathcal{C}\|\psi_n\|^{s^*}.
$$

If $m_n = 1$ then $sC_0\|\psi_n\|^{s(q^*-1)-q^*} \leq 1$, then

$$
-m_n \cdot \|\psi_n\|^{q^*} + m_n^s \cdot \|\psi_n\|^{s(q^*-1)} \cdot C_0 \leq -\mathcal{C}\|\psi_n\|^{q^*}.
$$

Altogether, we have that

$$
\mathrm{T}_\alpha(x_{n+1}) \leq \mathrm{T}_\alpha(x_n) - \mathcal{C} \cdot \min\{\|\psi_n\|^{s^*}, \|\psi_n\|^{q^*}\}.
$$

Since T_α is bounded from below we conclude that

$$
\|\psi_n\| \longrightarrow 0 \quad \text{as} \quad n \to \infty.
$$

Therefore, for all sufficiently big n we have

$$
\|\psi_n\| \leq 1.
$$

Then, we get that for all sufficiently big n we have

$$
\mathrm{T}_\alpha(x_{n+1}) \leq \mathrm{T}_\alpha(x_n) - \mathcal{C} \cdot \|\psi_n\|^{s^*}, \tag{4.14}
$$

since $s^* \geq q^*$. We introduce the numbers

$$
r_n := \mathrm{T}_\alpha(x_n) - \mathrm{T}_\alpha(x_\alpha^\delta),
$$

where x_α^δ is the minimizer of the Tikhonov functional. Due to the assumptions on X and Y we know that x_α^δ is unique, and therefore the numbers r_n are well-defined. We also remark that the numbers r_n may be regarded as a Bregman distance with respect to the functional T_α defined via

$$D_{\psi_\alpha^\delta}(x_\alpha^\delta, x_n) := T_\alpha(x_n) - T_\alpha(x_\alpha^\delta) - \langle \psi_\alpha^\delta, x_n - x_\alpha^\delta \rangle,$$

since x_α^δ is the minimizer of the Tikhonov functional the gradient of T_α at x_α^δ, which we denoted by ψ_α^δ, vanishes.

The next step of the proof is to connect r_n to $\|\psi_n\|$. The space X is c-convex. By Theorem 2.32, we have

$$\langle J_q^X(x_n) - J_q^X(x_\alpha^\delta), x_n - x_\alpha^\delta \rangle \geq C(\max\{\|x_n\|, \|x_\alpha^\delta\|\})^{q-c}\|x_n - x_\alpha^\delta\|^c.$$

By Theorem 2.41 and Theorem 2.42, we have $q \leq c$. Therefore, since x_n are uniformly bounded, we get

$$\langle J_q^X(x_n) - J_q^X(x_\alpha^\delta), x_n - x_\alpha^\delta \rangle \geq C\|x_n - x_\alpha^\delta\|^c.$$

Since $\psi_\alpha^\delta = \nabla T_\alpha(x_\alpha^\delta) = 0$ and since the subgradient of an convex functional is monotone, cf. Theorem 2.23, we have

$$\|\psi_n\|\|x_n - x_\alpha^\delta\| \geq \langle \psi_n - \psi_\alpha^\delta, x_n - x_\alpha^\delta \rangle$$
$$\geq \alpha\langle J_q^X(x_n) - J_q^X(x_\alpha^\delta), x_n - x_\alpha^\delta \rangle \geq C\|x_n - x_\alpha^\delta\|^c.$$

Hence,

$$\|\psi_n\| \geq C\|x_n - x_\alpha^\delta\|^{c-1}.$$

The above inequality proves that

$$\|x_n - x_\alpha^\delta\| \longrightarrow 0 \qquad \text{as} \quad n \to \infty,$$

since we have already proven that $\|\psi_n\| \to 0$ as $n \to \infty$ and since $c - 1 > 0$. Therefore, for all sufficiently big n we have

$$\|x_n - x_\alpha^\delta\| \leq 1.$$

We already know that

$$\|x_n - x_\alpha^\delta\| \leq C\|\psi_n\|^{c^*-1}.$$

Due to the definition of the subgradient, we have

$$r_n \leq \langle \psi_n, x_n - x_\alpha^\delta \rangle \leq \|\psi_n\|\|x_n - x_\alpha^\delta\|.$$

Hence, we get

$$r_n \leq C\|\psi_n\| \cdot \|\psi_n\|^{c^*-1} = C \cdot \|\psi_n\|^{c^*}.$$

By (4.14) we get that for all sufficiently big n

$$
\begin{aligned}
r_{n+1} &\leq r_n - C \cdot \|\psi_n\|^{s^*} \\
&\leq r_n - C \cdot r_n^{s^*/c^*}.
\end{aligned}
$$

Hence,

$$ r_n \longrightarrow 0 \qquad \text{as} \quad n \to \infty. $$

Next, we use the trick of Dunn [20, 19], which is common in the literature for convergence rates in Banach spaces, cf. e.g. [8, 43]. We set

$$ a := (s^*/c^*) - 1. $$

By the mean value theorem we have

$$ \frac{1}{r_{n+1}^a} - \frac{1}{r_n^a} = -a \frac{1}{\rho^{a+1}} (r_{n+1} - r_n) $$

with $\rho \in (r_{n+1}, r_n)$. Therefore, for all sufficiently big n we have

$$ \frac{1}{r_{n+1}^a} - \frac{1}{r_n^a} \geq C > 0. $$

Consequently, for all sufficiently big n and N we have

$$ \frac{1}{r_n^a} \geq \frac{1}{r_n^a} - \frac{1}{r_N^a} = \sum_{k=N}^{n-1} \frac{1}{r_{k+1}^a} - \frac{1}{r_k^a} \geq C \cdot (n - N). $$

Hence, for all $n \geq 0$ we get that

$$ r_n \leq C \cdot n^{-1/a}. $$

Since the functional $\frac{1}{p}\|Ax - y^\delta\|^p$ is convex and $\frac{1}{q}\|x\|^q$ is c-convex, we get for the Tikhonov functional T_α as the sum of the two aforementioned functionals that

$$ r_n \geq C \|x_n - x_\alpha^\delta\|^c. $$

For a proof of the above claim cf. Remark 4.19. Finally

$$ \|x_n - x_\alpha^\delta\| \leq C \cdot n^{-1/(a \cdot c)}, $$

which finishes the proof. $\qquad \square$

Remark 4.19. *In the proof above we claim that since the functional $\frac{1}{p}\|Ax - y^\delta\|^p$ is convex and $\frac{1}{q}\|x\|^q$ is c-convex, the Tikhonov functional T_α as the sum of the two aforementioned functionals is c-convex too in the sense that*

$$ \mathrm{T}_\alpha(x_n) - \mathrm{T}_\alpha(x_\alpha^\delta) = r_n \geq C \|x_n - x_\alpha^\delta\|^c. $$

This seems intuitively reasonable since it is common knowledge that the sum of two convex functionals always inherits the best of both convexity properties. A proof can be carried in the following way: Since X is c-convex we get by Corollary 2.49

$$\begin{aligned} D_{j_q}(x_\alpha^\delta, x_n) &= \tfrac{1}{q}\|x_n\|^q - \tfrac{1}{q}\|x_\alpha^\delta\|^q - \langle j_q(x_\alpha^\delta), x_n - x_\alpha^\delta \rangle \\ &\geq C \cdot (\|x_\alpha^\delta\| + \|x_n\|)^{q-c}\|x_\alpha^\delta - x_n\|^c \end{aligned}$$

resp.

$$\tfrac{1}{q}\|x_n\|^q \geq \tfrac{1}{q}\|x_\alpha^\delta\|^q + \langle j_q(x_\alpha^\delta), x_n - x_\alpha^\delta \rangle + C \cdot (\|x_\alpha^\delta\| + \|x_n\|)^{q-c}\|x_\alpha^\delta - x_n\|^c.$$

Further, we have

$$\tfrac{1}{p}\|Ax_n - y^\delta\|^p \geq \tfrac{1}{p}\|Ax_\alpha^\delta - y^\delta\|^p + \langle A^* j_p(Ax_\alpha^\delta - y^\delta), x_n - x_\alpha^\delta \rangle.$$

Multiplying the second last inequality by α and summing up with the last inequality we get

$$T_\alpha(x_n) \geq T_\alpha(x_\alpha^\delta) + \langle \psi_\alpha^\delta, x_n - x_\alpha^\delta \rangle + C \cdot (\|x_\alpha^\delta\| + \|x_n\|)^{q-c}\|x_\alpha^\delta - x_n\|^c,$$

where $\psi_\alpha^\delta \in \partial T_\alpha(x_\alpha^\delta)$. Since in the last theorem we also assumed that both X and Y are smooth spaces we also get that the subgradient is in fact a gradient. Therefore since x_α^δ is the minimizer of the Tikhonov functional we have $\psi_\alpha^\delta = 0$. Altogether we get

$$r_n \geq C \cdot (\|x_\alpha^\delta\| + \|x_n\|)^{q-c}\|x_\alpha^\delta - x_n\|^c.$$

Since if the space X is q-smooth and c-convex then $q \leq 2$ and $c \geq 2$. Hence

$$q - c \leq 0.$$

Finally since the iterates are bounded we get the desired claim.

The above theorem is not applicable if

$$s := \min\{p, q\} = c.$$

Since $p \leq 2, q \leq 2$ and $c \geq 2$, cf. Theorems 2.42 and 2.41, the above equality is only true if

$$p = q = c = 2.$$

In such case stronger convergence results can be established:

Theorem 4.20. *Let X be an 2-smooth, 2-convex Banach space and Y an 2-smooth Banach space and $p = q = 2$. Then,*

$$\|x_n - x_\alpha^\delta\| \leq C \cdot \exp(-n/C).$$

Proof. As in the proof of last theorem, we arrive at the inequality

$$T_\alpha(x_n - \mu\psi_n) \leq T_\alpha(x_n) - \left(\mu - \tfrac{1}{2}|\mu|^2 \left[G_Y\|A\|^2 + \alpha G_X\right]\right) \cdot \|\psi_n\|^2$$
$$\leq T_\alpha(x_n) - (\mu - \tfrac{1}{2}|\mu|^2 \cdot C_0) \cdot \|\psi_n\|^2$$

which is an analogue to the equation (4.12). The right-hand side of the last inequality is minimal if μ chosen as

$$m_n = 1/C_0.$$

Hence,

$$T_\alpha(x_{n+1}) \leq T_\alpha(x_n - m_n\psi_n) \leq T_\alpha(x_n) - C \cdot \|\psi_n\|^2.$$

Since X is 2-smooth and 2-convex we get that

$$\|\psi_n\|\|x_n - x_\alpha^\delta\| \geq \alpha\langle J_2(x_n) - J_2(x_\alpha^\delta), x_n - x_\alpha^\delta\rangle \geq C\|x_n - x_\alpha^\delta\|^2.$$

Therefore, we get that

$$\|x_n - x_\alpha^\delta\| \leq C\|\psi_n\|.$$

With $r_n := T_\alpha(x_n) - T_\alpha(x_\alpha^\delta)$ we conclude that

$$r_n \leq \langle \psi_n, x_n - x_\alpha^\delta\rangle \leq \|\psi_n\|\|x_n - x_\alpha^\delta\| \leq C\|\psi_n\|^2.$$

Therefore, we have

$$r_{n+1} \leq r_n - C \cdot \|\psi_n\|^2 \leq (1 - C)r_n.$$

Since the functional $\frac{1}{p}\|Ax - y^\delta\|^p$ is convex and $\frac{1}{q}\|x\|^q$ is 2-convex we get for the Tikhonov functional T_α as the sum of the two aforementioned functionals that

$$r_n \geq C\|x_n - x_\alpha^\delta\|^2,$$

which proves the claim, since the numbers r_n vanish with geometrical order. $\qquad\square$

4.4.2 Update in the dual space

In this section, we still consider minimizing the functional

$$T_\alpha(x) := \tfrac{1}{p}\|Ax - y^\delta\|_Y^p + \alpha \cdot \tfrac{1}{q}\|x\|_X^q \tag{4.15}$$

where throughout this section we assume that

$$p > 1 \quad \text{and} \quad X \text{ is } q\text{-convex}$$

and smooth of power type. But we make *no* assumptions on Y.

In [7] the authors have shown that the iteration defined via

$$x_{n+1}^* := x_n^* - \mu_n\psi_n \text{ with } \psi_n \in \partial T_\alpha(x_n) = A^* J_p^Y(Ax_n - y^\delta) + \alpha J_q^X(x_n),$$
$$x_{n+1} := J_{q^*}^{X^*}(x_{n+1}^*), \tag{4.16}$$

converges strongly to the unique minimizer x_α^δ of the Tikhonov functional T_α if the step size μ_n is chosen appropriately. In this section, we will introduce a highly improved version of the step size μ_n.

We stress that the choice of μ_n is crucial for the convergence properties of the steepest descent iteration. Alber, Iusem and Solodov [2] considered step sizes $\mu_n = \alpha_n \cdot (\|\partial T_\alpha(x_n)\|)^{-1}$ where the sequence $\{\alpha_n\}$ is chosen a priori and constrained only by the properties of X^*. They show that the iteration has a weak accumulation point in that case. In [7] the authors propose a step size selection strategy based on some (more or less) arbitrary chosen zero sequence. The authors are able to show strong convergence, however, due to the very general choice of the zero sequence the algorithm of [7] converges in general very slowly.

In this article we propose the selection of the step sizes μ_n as functions of the subdifferential of the Tikhonov functional T_α and some iteratively updated estimation of the Bregman distance. This unusual approach will prove to be very powerful. It will allow us to show convergence rates for the proposed method. In fact, we will show that although the convergence rates depend strongly on the properties of the underlying spaces X and Y; the algorithm needs only the information about the structure of X. The information about Y will be extracted from the subdifferential of T_α on-the-fly.

We start with the construction of the step size μ_n. Since X is convex of power type and smooth of power type, we can apply Theorem 2.52 and get

$$D_{j_q}(J_{q^*}(x_n^* - \mu\psi_n), x_\alpha^\delta)$$
$$= D_{j_q}(x_n, x_\alpha^\delta) + D_{j_{q^*}}(x_n^*, x_n^* - \mu\psi_n) + \mu\langle\psi_n, x_\alpha^\delta - x_n\rangle.$$

Since x_α^δ is the minimizer of T_α, there exists a single valued selection j_p such that

$$\psi_\alpha^\delta := 0 = A^* j_p(Ax_\alpha^\delta - y^\delta) + \alpha j_q(x_\alpha^\delta). \tag{4.17}$$

Therefore, with the monotonicity of the duality mapping, cf. Theorem 2.23, we get

$$\langle\psi_n, x_\alpha^\delta - x_n\rangle = \langle\psi_n - \psi_\alpha^\delta, x_\alpha^\delta - x_n\rangle \leq -\alpha\langle j_q(x_n) - j_q(x_\alpha^\delta), x_n - x_\alpha^\delta\rangle.$$

Further, we recall that

$$\langle j_q(x_n) - j_q(x_\alpha^\delta), x_n - x_\alpha^\delta\rangle = D_{j_q}(x_n, x_\alpha^\delta) + D_{j_q}(x_\alpha^\delta, x_n).$$

Hence, we have

$$-\mu\langle\psi_n, x_\alpha^\delta - x_n\rangle \leq -\mu\alpha \cdot D_{j_q}(x_n, x_\alpha^\delta).$$

Altogether, we get

$$D_{j_q}(J_{q^*}(x_n^* - \mu\psi_n), x_\alpha^\delta) \leq (1 - \mu\alpha)D_{j_q}(x_n, x_\alpha^\delta) + D_{j_{q^*}}(x_n^*, x_n^* - \mu\psi_n).$$

Of course we do not know the exact value of $D_{j_q}(x_n, x_\alpha^\delta)$. Assume that we know an upper estimate R_n of $D_{j_q}(x_n, x_\alpha^\delta)$, i.e.

$$D_{j_q}(x_n, x_\alpha^\delta) \leq R_n.$$

We arrive at the important inequality

$$D_{j_q}(J_{q^*}(x_n^* - \mu\psi_n), x_\alpha^\delta) \leq \max\{1 - \mu\alpha, 0\}R_n + D_{j_{q^*}}(x_n^*, x_n^* - \mu\psi_n).$$

The space X^* is q^*-smooth since X is q-convex, cf. Theorem 2.43. Therefore, there exists a constant $G_{q^*} \geq 0$ such that

$$D_{j_{q^*}}(x_n^*, x_n^* - \mu\psi_n) \leq |\mu|^{q^*} \cdot \frac{G_{q^*}}{q^*}\|\psi_n\|^{q^*}.$$

Finally, we arrive at the main inequality of our convergence analysis

$$D_{j_q}(J_{q^*}(x_n^* - \mu\psi_n), x_\alpha^\delta) \leq \max\{1 - \mu\alpha, 0\}R_n + |\mu|^{q^*}\frac{G_{q^*}}{q^*} \cdot \|\psi_n\|^{q^*}.$$

The optimal value of μ_n with respect to the right-hand side of the above inequality is given by

$$\mu_n = \min\{\left(\frac{\alpha}{G_{q^*}} \cdot \frac{R_n}{\|\psi_n\|^p}\right)^{\frac{1}{q^*-1}}, \frac{1}{\alpha}\}.$$

Now, we are ready to state the whole algorithm:

Algorithm 4.21. *Let G_{q^*} be the constant in the definition of smoothness of power type for the space X^*.*

(S_0) *Choose an arbitrary initial point $x_0 \in X$, dual initial point $x_0^* = J_q(x_0)$ and R_0 such that the condition $D_{j_q}(x_0, x_\alpha^\delta) \leq R_0$ is fulfilled. Set $n = 0$.*

(S_1) *Stop if $0 \in \partial T_\alpha(x_n)$, else choose $\psi_n \in \partial T_\alpha(x_n)$ and set*

$$\mu_n := \min\{\left(\frac{\alpha}{G_{q^*}} \cdot \frac{R_n}{\|\psi_n\|^{q^*}}\right)^{\frac{1}{q^*-1}}, \frac{1}{\alpha}\} \tag{4.18}$$

and

$$R_{n+1} := (1 - \mu_n\alpha)R_n + \mu_n^{q^*}\frac{G_{q^*}}{q^*} \cdot \|\psi_n\|^{q^*}.$$

(S_3) *Set*

$$x_{n+1}^* := x_n^* - \mu_n\psi_n \quad \text{and} \quad x_{n+1} := J_{q^*}^{X^*}(x_{n+1}^*).$$

(S_4) *Let $n \leftarrow (n+1)$ and go to step (S_1).*

Remark 4.22. *We remark that all the convergence results presented below also hold if the step size μ_n is alternatively chosen as*

$$\mu_n := \operatorname{argmin}_\mu \{\max\{1 - \mu\alpha, 0\} R_n + D_{j_{q^*}}(x_n^*, x_n^* - \mu\psi_n)\}$$

and R_{n+1} is updated by

$$R_{n+1} := (1 - \mu_n\alpha) R_n + D_{j_{q^*}}(x_n^*, x_n^* - \mu_n\psi_n).$$

Next, we prove the convergence rates results for the above algorithm. We introduce the auxiliary variable B_n defined via

$$B_n := \left(\frac{\alpha}{G_{q^*}} \cdot \frac{R_n}{\|\psi_n\|^{q^*}} \right)^{\frac{1}{q^*-1}}.$$

Then,

$$\mu_n = \min\{B_n, \frac{1}{\alpha}\} \tag{4.19}$$

and

$$R_{n+1} = [1 - \mu_n\alpha + \frac{\alpha}{q^*}\mu_n^{q^*} B_n^{-(q^*-1)}] R_n. \tag{4.20}$$

Theorem 4.23. *Let X be a q-convex Banach space and Y an arbitrary Banach space, then*

$$\|x_n - x_\alpha^\delta\| \leq C \cdot n^{-\frac{1}{q(q-1)}}.$$

Proof. We again use Dunn's trick to prove the convergence rate, cf. proof of Theorem 4.18. We have

$$\frac{1}{R_{n+1}^{q-1}} - \frac{1}{R_n^{q-1}} \geq \frac{1 - \left[1 - \mu_n\alpha + \frac{\alpha}{q^*}\mu_n^{q^*} B_n^{-(q^*-1)}\right]^{q-1}}{\left[1 - \mu_n\alpha + \frac{\alpha}{q^*}\mu_n^{q^*} B_n^{-(q^*-1)}\right]^{q-1}} \cdot \frac{1}{R_n^{q-1}}.$$

Since $0 < \mu_n\alpha - \frac{\alpha}{q^*}\mu_n^{q^*} B_n^{-(q^*-1)} \leq 1$ and $(1 - (1 - x)^y)/(1 - x)^y \geq yx$ for all $0 \leq x \leq 1$ and $y \geq 0$, we get

$$\frac{1}{R_{n+1}^{q-1}} - \frac{1}{R_n^{q-1}} \geq (q-1) \left(\mu_n\alpha - \frac{\alpha}{q^*}\mu_n^{q^*} B_n^{-(q^*-1)} \right) \cdot \frac{1}{R_n^{q-1}}.$$

If $\mu_n = B_n$, then

$$\mu_n\alpha - \frac{\alpha}{q^*}\mu_n^{q^*} B_n^{-(q^*-1)} = \frac{1}{q}\alpha B_n.$$

If on the other hand $\mu_n = \frac{1}{\alpha}$, then $B_n\alpha \geq 1$ and

$$\mu_n\alpha - \frac{\alpha}{q^*}\mu_n^{q^*} B_n^{-(q^*-1)} = 1 - \frac{1}{q^*}(B_n\alpha)^{-(q^*-1)} \geq 1 - \frac{1}{q^*} = \frac{1}{q}.$$

Hence,

$$\frac{1}{R_{n+1}^{q-1}} - \frac{1}{R_n^{q-1}} \geq \mathcal{C} \cdot \min\{B_n, 1\} \frac{1}{R_n^{q-1}}.$$

Next, we will show that the right-hand side is uniformly bounded away from zero. By construction, we have $1/R_n^{q-1} \geq 1/R_0^{q-1}$. Therefore, the numbers $1/R_n^{q-1}$ are uniformly bounded away from zero. Next, we consider the numbers B_n/R_n^{q-1}. We have

$$B_n/R_n^{q-1} = \left(\frac{\alpha}{G_{q^*}} \cdot \frac{1}{\|\psi_n\|^{q^*}}\right)^{\frac{1}{q^*-1}}.$$

To estimate $(\|\psi_n\|^{q^*})$, we need the following lemma:

Lemma 4.24. *Let X be a normed space and $f : X \to \mathbb{R}$ a convex function. If $|f|$ is bounded on all bounded sets, then also ∂f is bounded on all bounded sets.*

Proof. Let $A \subset B(0, R)$ be an arbitrary bounded set. Then,

$$F := \sup\{|f(x)| : x \in B(0, R+1)\} < \infty.$$

Assume $x \in A, \psi \in \partial f(x), y \in X$ with $\|y\| = 1$. Then, $x + y \operatorname{sign}\langle\psi, y\rangle \in B(0, R+1)$. By definition of the subgradient, we get

$$|\langle\psi, y\rangle| = \langle\psi, x + y \operatorname{sign}\langle\psi, y\rangle - x\rangle \leq f(x + y \operatorname{sign}\langle\psi, y\rangle) - f(x) \leq 2F.$$

Hence, $\sup\{\|\psi\| : \psi \in \partial f(x) \text{ with } x \in A\} \leq 2F < \infty$. $\qquad\square$

By construction, the sequence $D_{j_q}(x_n, x_\alpha^\delta)$ is bounded. Therefore, by Theorem 2.48, the sequence (x_n) is also bounded. Finally, by Lemma 4.24, the sequence (ψ_n) is bounded, since any norm is bounded on bounded sets. Therefore, the numbers B_n/R_n^{q-1} are uniformly bounded away from zero. Thus, $1/R_{n+1}^{q-1} - 1/R_n^{q-1} \geq \mathcal{C}$ uniformly in n. Therefore, we have

$$\frac{1}{R_{n+1}^{q-1}} \geq \frac{1}{R_{n+1}^{q-1}} - \frac{1}{R_0^{q-1}} = \sum_{k=0}^{n} \frac{1}{R_{k+1}^{q-1}} - \frac{1}{R_k^{q-1}} \geq (n+1)\mathcal{C}.$$

We conclude that

$$R_{n+1}^{q-1} \leq \mathcal{C}(n+1)^{-1}.$$

Since X is q-convex we have, cf. Theorem 2.48,

$$\|x_n - x_\alpha^\delta\|^q \leq \mathcal{C} \cdot D_{j_q}(x_n, x_\alpha^\delta) \leq \mathcal{C} \cdot R_n \leq \mathcal{C} \cdot n^{-1/(q-1)}.$$

Which proves the claim. $\qquad\square$

Theorem 4.25. *Assume that additionally to the assumptions made in Theorem 4.23 the space Y is p-convex and*

$$p > 2 \quad \text{or} \quad q > 2,$$

then the convergence rate improves to

$$\|x_n - x_\alpha^\delta\| \leq C \cdot n^{-\frac{(M-1)}{[(M-1)(q-1)-1]q}},$$

where

$$M := \max\{p, q\} > 2.$$

Proof. We can carry out the proof along the lines of the proof of Theorem 4.23. The main difference is our estimation of the numbers B_n. First, we need the following lemma:

Lemma 4.26. *Let X be q-convex and Y be p-convex where $q > 2$ or $p > 2$. Then, the subdifferential of the Tikhonov functional T_α is locally γ-Hölder continuous, where*

$$\gamma = \min\left\{ \frac{1}{(p-1)}, \frac{1}{(q-1)} \right\}.$$

Proof. Let B be a bounded set and $x, y \in B$. Then,

$$\|\partial \mathrm{T}_\alpha(x) - \partial \mathrm{T}_\alpha(y)\| \leq \|A^*\| \|j_p(Ax - y^\delta) - j_p(Ay - y^\delta)\| + \alpha \|j_q(x) - j_q(y)\|.$$

Since X is q-convex, the dual space X^* is q^*-smooth, cf. Theorem 2.43. Further, $q^* \leq q$ by Theorems 2.41 and 2.42. Therefore, we get by Theorem 2.33

$$\|j_q(x) - j_q(y)\| \leq C(\max\{\|x\|, \|y\|\})^{q-q^*} \cdot \|x - y\|^{q^*-1} \leq C\|x - y\|^{q^*-1}$$

and

$$\|j_p(Ax - y^\delta) - j_p(Ay - y^\delta)\| \leq C\|Ax - Ay\|^{p^*-1} \leq C\|x - y\|^{p^*-1}.$$

Since $q^* - 1 = \frac{1}{q-1}$ and $p^* - 1 = \frac{1}{p-1}$ we get

$$\|\partial \mathrm{T}_\alpha(x) - \partial \mathrm{T}_\alpha(y)\| \leq C \cdot \left(\|x - y\|^{\frac{1}{q-1}} + \|x - y\|^{\frac{1}{p-1}} \right) \leq C\|x - y\|^\gamma,$$

which proves the claim. $\qquad\square$

We recall that the iterates x_n are uniformly bounded. With $M := \max\{p, q\}$ and (4.17) we can estimate

$$\|\psi_n\|^{q^*} \leq C \cdot \|x_n - x_\alpha^\delta\|^{q\frac{q^*}{q(M-1)}} \leq C \cdot R_n^{\frac{q^*}{q(M-1)}} = C \cdot R_n^{\frac{1}{(M-1)(q-1)}},$$

since X is q-convex. We define the auxiliary variable γ via

$$\gamma := \frac{1}{(M-1)(q-1)} < 1.$$

As in the proof of Theorem 4.23, we can estimate

$$B_n / R_n^{(1-\gamma)(q-1)} \geq C \cdot \left(\frac{\alpha}{G_{q^*}} \cdot \frac{R_n}{R_n^\gamma} \right)^{\frac{1}{q^*-1}} R_n^{(1-\gamma)(q-1)} \geq C > 0.$$

Then, the difference $1/R_{n+1}^{(1-\gamma)(q-1)} - 1/R_n^{(1-\gamma)(q-1)}$ is uniformly bounded away from zero. Therefore, we have $R_n^{(1-\gamma)(q-1)} \leq C \cdot n^{-1}$. Hence, we get

$$\|x_n - x_\alpha^\delta\| \leq C \cdot n^{-\frac{1}{(1-\gamma)(q-1)q}}.$$

with

$$\frac{1}{1-\gamma} = \frac{(M-1)(q-1)}{(M-1)(q-1)-1} > 0,$$

which proves the claim. $\square$

Theorem 4.27. *Let X and Y be 2-convex and $p = q = 2$. Then, there exists a constant $C > 0$, such that*

$$\|x_n - x_\alpha^\delta\| \leq C \cdot \exp(-n/C).$$

Proof. As in the proofs of Theorem 4.25 and Theorem 4.23, the main key to the desired convergence rate is the estimation of $\|\psi_n\|$. By the same technique as in proof of lemma 4.26 (and therein employing Theorem 2.33) we get that the subdifferential ∂T_α is locally Lipschitz continuous, therefore

$$\|\psi_n\|^2 \leq C \cdot \|x_n - x_\alpha^\delta\|^2 \leq C \cdot R_n.$$

Hence, the numbers B_n are uniformly bounded away from zero. By (4.19) and (4.20) there exists a $0 \leq \gamma < 1$, such that $R_{n+1} \leq \gamma \cdot R_n \leq \gamma^n R_0$ for all n. Therefore, the sequence (R_n) vanishes geometrically. But the sequence (R_n) dominates $(\|x_n - x_\alpha^\delta\|^2)$ up to some factor. This proves the claim. $\square$

5

Landweber regularization

The aim of this section is to show how some of the results known for the Landweber iteration in Hilbert spaces can be translated to the setting of Banach spaces.

First, we recall some facts about Landweber iteration in Hilbert spaces. We want to regularize the problem $Ax = y$ with $A : X \to Y$, where X and Y are Hilbert spaces. The Landweber iteration is given by

$$x_{n+1} = x_n - \mu_n A^*(Ax_n - y^\delta), \tag{5.1}$$

for μ_n chosen appropriately. Since $A^*(Ax_n - y^\delta) = (\frac{1}{2}\|A \cdot -y^\delta\|^2)'(x_n)$ the above iteration can be interpreted as a steepest descent method for the functional

$$\tfrac{1}{2}\|Ax - y^\delta\|^2.$$

Since the problem $Ax = y$ is ill-posed, it is clear that the the steepest descent has to be stopped at an appropriate step, say $N(\delta, y^\delta)$, to ensure the regularization, i.e. $x^\delta_{N(\delta, y^\delta)} \to x^\dagger$ as $\delta \to 0$.

A generalization of the iteration (5.1) to the setting of Banach spaces has been introduced in [59]. It is given by

$$x^*_{n+1} = x^*_n - \mu_n \psi_n \qquad \psi_n = A^* j^Y_p(Ax_n - y^\delta) \in \partial(\tfrac{1}{p}\|A \cdot -y^\delta\|^p)(x_n)$$

$$x_{n+1} = J^{X^*}_{q^*}(x^*_{n+1}).$$

Let (y_k) be a sequence of noisy data with $\|y_k - y\| \le \delta_k$, where the sequence (δ_k) is non-increasing and $\delta_k \to 0$ as $k \to \infty$. Then, a sequence (x_n) is generated via the above method as long as $\|Ax_n - y_k\| > \tau \delta_k$ for some a-priori chosen $\tau > 1$ and the last iterate of the noise level δ_k is used as the start iterate for the noise level δ_{k+1}.

If the step sizes μ_n are chosen appropriately, then one can show that the resulting sequence (x_n) is a regularizing sequence, i.e. $x_n \to x^\dagger$ as $n \to \infty$, where $x^\dagger$ is the minimal norm solution of $Ax = y$.

We remark two things about the above iteration: First, we notice that for fixed noise level the discrepancy principle of Morozov is used to stop the iteration. Second, we remark that a change of a single y_k in the sequence of noisy data (y_k) results in a completely new regularizing sequence, which has only finitely many common elements with the original sequence. In this sense the above regularizing result is a sequence-to-sequence regularizing result. This is a main difference to the definition of regularization considered in the literature, where for every element of the (y_k) an element of the sequence (x_n) is generated, cf. Definition 2.56. In contrast to sequence-to-sequence regularization we call such results element-to-element regularization. Therefore, the change of a single y_k in the sequence of noisy data (y_k) results in a change of only a single element of the sequence (x_n).

In this section, we show the following results

1. The algorithm of [59] was designed for the case of X being uniformly convex. Therefore, we discuss the improvements for the case that X is not only uniformly convex but convex of power type. This will especially affect the construction of the step sizes μ_n. We will show that for the improved choice of μ_n an analogue to the original sequence-to-sequence regularization results holds.

2. Our second aim in this section is to show element-to-element regularization results and in particular to obtain convergence rates. To do so we will have to modify the iteration. We will show that for a modified version of the Landweber iteration defined by

$$x_{n+1}^* = x_n^* - \mu_n \psi_n - \beta_n x_n^*$$
$$\psi_n = A^* j_p^Y (Ax_n - y^\delta) \in \partial(\tfrac{1}{p}\|A \cdot -y^\delta\|^p)(x_n)$$
$$x_{n+1} = J_{q^*}^{X^*}(x_{n+1}^*),$$

where μ_n and β_n are chosen appropriately a convergence rate of the form

$$D_{j_q}(x_{N(\delta,y^\delta)}, x^\delta) \leq C \cdot \delta$$

can be shown under the low order source condition $j_q(x^\dagger) = A^*\omega$. It turns out that to show the convergence rates we will not only have to modify the iteration, but also the stopping criterion. The new stopping criterion will be based on the properties of the step size β_n. We remark that although this might seem unusual at the first glance, it is not an uncommon way to achieve convergence rates. A similar stopping criterion was also used e.g. by [44].

5.1   Landweber regularization

The aim of this section is to provide an appropriate choice for the step sizes μ_n of the iteration

$$
\begin{aligned}
x_{n+1}^* &:= x_n^* - \mu_n \psi_n \qquad \psi_n := A^* j_p^Y (Ax_n - y^\delta) \in \partial(\tfrac{1}{p}\|A\cdot - y^\delta\|^p)(x_n) \\
x_{n+1} &:= J_{q^*}^{X^*}(x_{n+1}^*),
\end{aligned}
\tag{5.2}
$$

where throughout this section we assume that

$$ X \quad \text{is} \quad q\text{-convex} $$

and smooth of power type. The iteration (5.2) may be regarded as a steepest descent in the dual space for the functional

$$ \tfrac{1}{p}\|Ax - y^\delta\|^p. $$

As we have seen in Section 4.4.2 the choice of the step size μ_n is crucial for the convergence properties of methods in the dual space.

We start with the following important lemma:

Lemma 5.1. *Let X be convex of power type q and smooth of power type. Further, let $\tau\delta \le \|Ax_n - y^\delta\|$ with $\tau > 1, \delta \ge 0$ and let the step size μ_n be given by*

$$
\mu_n := \left[(\|Ax_n - y^\delta\|^p - \delta\|Ax_n - y^\delta\|^{p-1})/(G_{q^*}\|\psi_n\|^{q^*}) \right]^{q-1},
$$

where G_{q^*} is the constant in the definition of smoothness of power type for the space X^*. Then, $\mu_n > 0$ and

$$
\begin{aligned}
D_{j_q}(x_{n+1}, x^\dagger) &\le D_{j_q}(x_n, x^\dagger) - C \cdot \mu_n \|Ax_n - y^\delta\|^p \\
&\le D_{j_q}(x_n, x^\dagger) - C \cdot \|Ax_n - y^\delta\|^q.
\end{aligned}
$$

Where C does not depend on n nor y^δ.

Proof. By the three point identity, cf. Lemma 2.52, and due to $x_n^* - x_{n+1}^* = \mu_n \psi_n$ we have

$$
D_{j_q}(x_{n+1}, x^\dagger) = D_{j_q}(x_n, x^\dagger) + D_{j_{q^*}}(x_n^*, x_{n+1}^*) + \langle \mu_n \psi_n, x^\dagger - x_n \rangle.
$$

By the properties of the duality mapping we get

$$
\begin{aligned}
\mu_n \langle \psi_n, x^\dagger - x_n \rangle &= -\mu_n \langle \psi_n, x_n - x^\dagger \rangle \\
&= -\mu_n \langle j_p(Ax_n - y^\delta), Ax_n - y^\delta + y^\delta - Ax^\dagger \rangle \\
&\le -\mu_n \|Ax_n - y^\delta\|^p + |\mu_n|\delta\|Ax_n - y^\delta\|^{p-1}.
\end{aligned}
$$

Since X is convex of power type q, the dual space X^* is smooth of power type q^* (cf. Theorem 2.43). Therefore, together with Lemma 2.48 we have

$$D_{j_{q^*}}(x_n^*, x_{n+1}^*) \leq |\mu_n|^{q^*} \frac{G_{q^*}}{q^*} \|\psi_n\|^{q^*}.$$

Altogether we get

$$\begin{aligned}
D_{j_q}(x_{n+1}, x^\dagger) &\leq D_{j_q}(x_n, x^\dagger) - \mu_n \|Ax_n - y^\delta\|^p \\
&\quad + |\mu_n| \delta \|Ax_n - y^\delta\|^{p-1} + |\mu_n|^{q^*} \frac{G_{q^*}}{q^*} \|\psi_n\|^{q^*}.
\end{aligned}$$

We notice that the step size μ_n is optimal with respect to the right-hand side of the above inequality. Since $\tau > 1$ and $\tau\delta \leq \|Ax_n - y^\delta\|$ we have that $\mu_n > 0$. Further, we have

$$\mu_n^{q^*} \frac{G_{q^*}}{q^*} \|\psi_n\|^{q^*} = \mu_n \|Ax_n - y^\delta\|^p - \mu_n \delta \|Ax_n - y^\delta\|^{p-1}.$$

Therefore, we get

$$\begin{aligned}
&-\mu_n \|Ax_n - y^\delta\|^p + |\mu_n| \delta \|Ax_n - y^\delta\|^{p-1} + |\mu_n|^{q^*} \frac{G_{q^*}}{q^*} \|\psi_n\|^{q^*} \\
&= -(1 - \tfrac{1}{q^*})(\mu_n \|Ax_n - y^\delta\|^p - \mu_n \delta \|Ax_n - y^\delta\|^{p-1}) \\
&\leq -(1 - \tfrac{1}{q^*})(1 - \tfrac{1}{\tau})\mu_n \|Ax_n - y^\delta\|^p.
\end{aligned}$$

This proves that

$$D_{j_q}(x_{n+1}, x^\dagger) \leq D_{j_q}(x_n, x^\dagger) - (1 - \tfrac{1}{q^*})(1 - \tfrac{1}{\tau})\mu_n \|Ax_n - y^\delta\|^p.$$

Due to $\psi_n = A^* j_p(Ax_n - y^\delta)$, the properties of the duality mapping and $\tau\delta \leq \|Ax_n - y^\delta\|$ we have

$$\mu_n \geq \mathcal{C} \|Ax_n - y^\delta\|^{q-p}.$$

Altogether we get

$$\begin{aligned}
D_{j_q}(x_{n+1}, x^\dagger) &\leq D_{j_q}(x_n, x^\dagger) - \mathcal{C}\mu_n \|Ax_n - y^\delta\|^p \\
&\leq D_{j_q}(x_n, x^\dagger) - \mathcal{C} \|Ax_n - y^\delta\|^q,
\end{aligned}$$

which proves the claim. $\square$

We can show the following regularization result:

Theorem 5.2. *Let X be a Banach space convex of power type q and smooth of power type, Y an arbitrary Banach space, $A : X \to Y$ a linear, continuous operator. Further, let (δ_k) be monotonically vanishing, i.e.*

$$\delta_{k+1} \leq \delta_k \quad \text{and} \quad \delta_k \longrightarrow 0.$$

and (y^{δ_k}) such that $\|y^{\delta_k} - y\| \leq \delta_k$. Further, let $\tau > 1, p > 1$ be fixed. Then, we generate a sequence (x_n) as follows:

(S_0) *Choose $x_0 \in X$ such that $x_0^* := j_q(x_0) \in \mathcal{R}(A^*)$. Notice that this is always possible with $x_0 = x_0^* = 0$. Set $k = 0$ and $n = 0$.*

(S_1) *STOP if*

$$\|Ax_n - y^{\delta_\kappa}\| \leq \tau \delta_\kappa$$

for all $\kappa \geq k$ else: if $\|Ax_n - y^{\delta_k}\| > \tau \delta_k$ go to step (S_2) else set $k + 1 \leftarrow k$ and restart step (S_1).

(S_2) *Set*

$$\mu_n := \left[\left(\|Ax_n - y^{\delta_k}\|^p - \delta \|Ax_n - y^{\delta_k}\|^{p-1} \right) / \left(G_{q^*} \|\psi_n\|^{q^*} \right) \right]^{q-1}$$

and

$$x_{n+1}^* := x_n^* - \mu_n \psi_n \qquad \psi_n := A^* j_p^Y (Ax_n - y^{\delta_k}) \in \partial(\tfrac{1}{p} \|A \cdot - y^{\delta_k}\|^p)(x_n)$$
$$x_{n+1} := J_{q^*}^{X^*}(x_{n+1}^*).$$

Set $n + 1 \leftarrow n$ and go to (S_1).

Then, either the iteration terminates after finite number of steps with the last iterate being the minimum norm solution $x^\dagger$ of $Ax = y$ or

$$\|x_n - x^\dagger\| \to 0 \qquad as \quad n \to \infty.$$

Proof. Assume that the iteration does not stop after finitely many steps. By Lemma 5.1 the distances $D_{j_q}(x_n, x^\dagger)$ are monotonically decreasing and bounded, hence convergent. In the next step we will show that

$$\lim_n D_{j_q}(x_n, x^\dagger) = 0.$$

By $k(n)$ we denote the value of k for in step (S_2). Therefore, we have $\|Ax_n - y^{\delta_{k(n)}}\| > \tau \delta_{k(n)}$. Then, due to Lemma 5.1 we have

$$\mathcal{C}\|Ax_n - y^{\delta_{k(n)}}\|^q \leq [D_{j_q}(x_n, x^\dagger) - D_{j_q}(x_{n+1}, x^\dagger)] \longrightarrow 0$$

for $n \to \infty$. Hence, also $\|Ax_n - y^{\delta_{k(n)}}\| \to 0$. Therefore, there exists a sub-sequence $(x_{n_m})_m$ such that for all m we have

$$\|Ax_{n_m} - y^{\delta_{k(n_m)}}\| \leq \|Ax_n - y^{\delta_{k(n)}}\| \qquad \text{for all} \quad n \leq n_m.$$

In particular this implies that the sequence $(\|Ax_{n_m} - y^{\delta_{k(n_m)}}\|)_m$ vanishes monotonically.

Let $\epsilon > 0$ be fixed. For $i < j$ we have

$$D_{j_q}(x_{n_i}, x_{n_j}) = D_{j_q}(x_{n_i}, x^\dagger) - D_{j_q}(x_{n_j}, x^\dagger) - \langle j_q(x_{n_j}) - j_q(x_{n_i}), x^\dagger - x_{n_j} \rangle$$

Hence, for sufficiently large i, j we have

$$D_{j_q}(x_{n_i}, x^\dagger) - D_{j_q}(x_{n_j}, x^\dagger) \leq \epsilon.$$

Since $j_q(x_n) = x_n^*$ we have $\langle j_q(x_{n_j}) - j_q(x_{n_i}), x^\dagger - x_{n_j} \rangle = \langle x_{n_j}^* - x_{n_i}^*, x^\dagger - x_{n_j} \rangle$.
Further, we have for $i < j$ that

$$-\langle x_{n_j}^* - x_{n_i}^*, x^\dagger - x_{n_j} \rangle = \sum_{n=n_i}^{n_j-1} \langle x_{n+1}^* - x_n^*, x_{n_j} - x^\dagger \rangle$$

with

$$
\begin{aligned}
\langle x_{n+1}^* &- x_n^*, x_{n_j} - x^\dagger \rangle \\
&= \langle -\mu_n A^* j_p(Ax_n - y^{\delta_{k(n)}}), x_{n_j} - x^\dagger \rangle \\
&= -\mu_n \langle j_p(Ax_n - y^{\delta_{k(n)}}), Ax_{n_j} - Ax^\dagger \rangle \\
&= -\mu_n \langle j_p(Ax_n - y^{\delta_{k(n)}}), Ax_{n_j} - y^{\delta_{k(n_j)}} + y^{\delta_{k(n_j)}} - Ax^\dagger \rangle \\
&\leq \mu_n \|Ax_n - y^{\delta_{k(n)}}\|^{p-1} \|Ax_{n_j} - y^{\delta_{k(n_j)}}\| + \mu_n \|Ax_n - y^{\delta_{k(n)}}\|^{p-1} \delta_{k(n_j)} \\
&\leq (1 + \tfrac{1}{\tau}) \mu_n \|Ax_n - y^{\delta_{k(n)}}\|^{p-1} \|Ax_{n_j} - y^{\delta_{k(n_j)}}\|.
\end{aligned}
$$

By assumption we have that $\|Ax_{n_j} - y^{\delta_{k(n_j)}}\| \leq \|Ax_n - y^{\delta_{k(n)}}\|$ for $n < n_j$. Therefore, we get

$$-\langle x_{n_j}^* - x_{n_i}^*, x^\dagger - x_{n_j} \rangle \leq (1 + \tfrac{1}{\tau}) \sum_{n=n_i}^{n_j-1} \mu_n \|Ax_n - y^{\delta_{k(n)}}\|^p.$$

Due to Lemma 5.1 we have

$$\sum_{n=0}^{\infty} \mu_n \|Ax_n - y^{\delta_{k(n)}}\|^p < \infty.$$

Therefore, for sufficiently large i, j, we have

$$-\langle x_{n_j}^* - x_{n_i}^*, x^\dagger - x_{n_j} \rangle \leq \epsilon.$$

Altogether we have shown that for every $\epsilon > 0$ we have

$$D_{j_q}(x_{n_i}, x_{n_j}) \leq 2\epsilon.$$

for all sufficiently large i and j. Therefore, by the convexity of power type q of X we have

$$C\|x_{n_i} - x_{n_j}\|^q \leq D_{j_q}(x_{n_i}, x_{n_j}) \leq 2\epsilon.$$

Hence, the sequence (x_{n_m}) is Cauchy. Since the space X is a Banach space the sequence (x_{n_m}) is also strongly convergent. Let x be the limit of (x_{n_m}), i.e.

$$x := \lim_m x_{n_m}.$$

We have

$$0 \leq \|Ax_{n_m} - y\| \leq \|Ax_{n_m} - y^{\delta_{k(n_m)}}\| + \|y^{\delta_{k(n_m)}} - y\|.$$

The left-hand side of the last inequality converges to $\|A(x - x^\dagger)\|$, whereas the right-hand side converges to zero. Altogether this proves that $\|A(x - x^\dagger)\| = 0$, or

$$x - x^\dagger \in \mathcal{N}(A).$$

Since by assumption $x_0^* \in \mathcal{R}(A^*)$ and $\psi_n \in \mathcal{R}(A^*)$ for all n, we have $x_n^* \in \mathcal{R}(A^*)$ for all n. Hence,

$$j_q(x) \in \overline{\mathcal{R}(A^*)}.$$

Due to Lemma 2.55 we have

$$x = x^\dagger.$$

Since X is convex of power type the mapping $D_{j_q}(\cdot, x^\dagger)$ is continuous, cf. Theorem 2.48. Hence, we have that

$$\lim_m D_{j_q}(x_{n_m}, x^\dagger) = D_{j_q}(x^\dagger, x^\dagger) = 0.$$

Since the sequence $(D_{j_q}(x_n, x^\dagger))$ is convergent and has a sub-sequence converging to zero, it converges to zero. Hence, by the convexity of power type of X we conclude that the sequence (x_n) converges strongly to the minimum norm solution $x^\dagger$.

Finally, we consider the case that the iteration stops after a finite number of steps. This is only the case if for the final iterate, say x_N, we have $\|Ax_N - y\| = 0$. Therefore, $x_N - x^\dagger \in \mathcal{N}(A)$. Further, by construction we have $j_q(x_N) = x_N^* \in \mathcal{R}(A^*)$. Hence, again by Lemma 2.55 we have $x_N = x^\dagger$. □

Remark 5.3. *In the above proof we need the explicit form of the step sizes μ_n only to show that*

$$\sum_{n=0}^{\infty} \mu_n \|Ax_n - y^{\delta_{k(n)}}\|^p < \infty.$$

Therefore, the above theorem also holds if the original choice of μ_n is replaced by

$$\mu_n^* = \operatorname*{argmin}_\mu \{-\mu\|Ax_n - y^{\delta_k}\|^p + |\mu|\delta\|Ax_n - y^{\delta_k}\|^{p-1} + D_{j_{q^*}}(x_n^*, x_n^* - \mu\psi_n)\}$$

$$\mu_n = \min\{\mu_n^*, \overline{\mu}\|Ax_n - y^{\delta_k}\|^{q-p}\},$$

for every fixed $\overline{\mu} > 0$. For a detailed proof, cf. [32].

5.2   Modified Landweber regularization

In the last section we have shown that the iteration (5.2) together with μ_n as in Lemma 5.1 may be used to generate a regularizing sequence, i.e. a sequence (x_n) such that $x_n \to x^\dagger$. Since we have not assumed any source conditions on the minimum norm solution $x^\dagger$, therefore we cannot hope to show convergence rates.

Therefore, the aim of this section is to introduce a version of the iteration (5.2), which is modified in such a way that a convergence rate can be shown if the low order source condition (4.6) is assumed.

To do so we will consider a particular version of the iteration

$$
\begin{aligned}
x^*_{n+1} &= x^*_n - \mu_n \psi_n - \beta_n x^*_n \\
\psi_n &= A^* j^Y_p (A x_n - y^\delta) \in \partial(\tfrac{1}{p} \| A \cdot -y^\delta \|^p)(x_n) \\
x_{n+1} &= J^{X^*}_{q^*}(x^*_{n+1})
\end{aligned}
\tag{5.3}
$$

We recall that the proofs of the theorems in the last section depended heavily on Lemma 5.1 and therefore on the right choice of the step size μ_n in iteration (5.2). In contrast to this approach, the iteration introduced in this section will employ the properties of the sequence β_n rather than the parameter μ_n.

We will also no longer use the discrepancy principle $\|A x_n - y^\delta\| \leq \tau \delta$ to stop the iteration but a stopping criterion inspired by [44]. Namely we will stop the iteration if $\tau \beta_n \leq \delta^{p-1}$.

Throughout this section, we will assume that

$$
X \quad \text{is} \quad p - \text{convex.}
$$

Hence, contrary to the original iteration (5.2) the indices of the duality mapping j^Y_p and the power of convexity of X are coupled.

First, we prove an analogue of Lemma 5.1:

Lemma 5.4. *Let X be a Banach space convex of power type p and j_p be chosen such that $j_p(x_{n+1}) = x^*_{n+1} = x^*_n - \mu_n \psi_n - \beta_n x^*_n$ with $\psi_n = A^* j^Y_p (A x_n - y^\delta)$, $j_p(x_n) = x^*_n$ and $j_p(x^\dagger) = A^* \omega$ with $\|\omega\| \leq S$. Then*

$$
\begin{aligned}
D_{j_p}&(x_{n+1}, x^\dagger) \\
&\leq (1 - \beta_n) D_{j_p}(x_n, x^\dagger) - \mu_n \|A x_n - y^\delta\|^p + |\mu_n| \delta \|A x_n - y^\delta\|^{p-1} \\
&\quad + S \|A x_n - y^\delta\| \beta_n + S \delta \beta_n + D_{j_{p^*}}(x^*_n, x^*_{n+1}).
\end{aligned}
$$

Proof. By the three point identity, cf. Lemma 2.52, we have

$$
D_{j_p}(x_{n+1}, x^\dagger) = D_{j_p}(x_n, x^\dagger) + D_{j_{p^*}}(x^*_n, x^*_{n+1}) + \langle \mu_n \psi_n + \beta_n x^*_n, x^\dagger - x_n \rangle.
$$

We can estimate the last term by

$$\langle \mu_n \psi_n, x^\dagger - x_n \rangle$$
$$= -\mu_n \langle j_p(Ax_n - y^\delta), Ax_n - y^\delta \rangle - \mu_n \langle j_p(Ax_n - y^\delta), y^\delta - y \rangle$$
$$\leq -\mu_n \|Ax_n - y^\delta\|^p + |\mu_n|\delta\|Ax_n - y^\delta\|^{p-1}.$$

and

$$\langle \beta_n x_n^*, x^\dagger - x_n \rangle$$
$$= -\beta_n \langle j_p(x^\dagger) - x_n^*, x^\dagger - x_n \rangle + \beta_n \langle j_p(x^\dagger), x^\dagger - x_n \rangle$$
$$= -\beta_n [D_{j_p}(x_n, x^\dagger) + D_{j_p}(x^\dagger, x_n)] + \beta_n \langle \omega, y - y^\delta + y^\delta - Ax_n \rangle$$
$$\leq -\beta_n D_{j_p}(x_n, x^\dagger) + |\beta_n| \cdot S \cdot \delta + |\beta_n| \cdot S\|Ax_n - y^\delta\|.$$

Which concludes the proof. □

Algorithm 5.5. *Let X be a Banach space convex of power type p, Y an arbitrary Banach space, $A : X \to Y$ linear, continuous operator. Let $\delta \geq 0$ and y^δ be such that $\|y^\delta - y\| \leq \delta$. Further, let $\tau > 0$ be fixed. Let $x^\dagger$ be the minimum norm solution of the problem $Ax = y$ and assume that the source condition*

$$A^*\omega = j_p^X(x^\dagger), \tag{5.4}$$

holds for some $\omega \in Y^*$ with $\|\omega\| \leq S$. Let G_{p^*} be the constant in the definition of smoothness of power type for X^*. Let the modified iteration be given via

(S_0) *Initialization. Choose $\mu > 0$ fixed such that*

$$(-\mu + \tfrac{1}{p^*}\mu^{p^*} + \tfrac{1}{p}\mu^p + 2^{p^*}\tfrac{G_{p^*}}{p^*}\|A\|^{p^*}\mu^{p^*}) \leq 0,$$

start point x_0^*, $x_0 := J_{p^*}(x_0^*)$, a number R_0 such that $D_{j_p}(x_0, x^\dagger) \leq R_0$. Set $n \leftarrow 0$.

(S_1) *STOP if $\tau \beta_n \leq \delta^{p-1}$, else set*

$$\gamma_n = \tfrac{p^*}{p}\tau^{p^*} + S^{p^*}\mu^{-p^*} + p^*S\tau^{p^*-1} + 2^{p^*-1}G_{p^*}\|x_n^*\|^{p^*}$$
$$\beta_n = \min\{1, (R_n/\gamma_n)^{p-1}\}$$
$$R_{n+1} = (1 - \beta_n)R_n + \tfrac{\gamma_n}{p^*}\beta_n^{p^*}$$

(S_2) *Compute the new iterate via*

$$x_{n+1}^* = x_n^* - \mu A^* j_p(Ax_n - y) - \beta_n x_n^*$$
$$x_{n+1} = J_{p^*}^{X^*}(x_{n+1}^*).$$

Set $n \leftarrow (n + 1)$ and go to step (S_1).

Theorem 5.6. *For $\delta > 0$ let $N(\delta, y^\delta)$ be the index where the modified iteration stops, then there exists a number $C > 0$ independent of δ (and y^δ) such that*

$$D_{j_p}(x_{N(\delta,y^\delta)}, x^\dagger) \le C \cdot \delta.$$

for all sufficiently small δ.

Proof. The proof will have two parts:

1. First we will show that the number $N(\delta, y^\delta)$ is well-defined.

2. Then, we will show the convergence rate.

Since X is convex of power type p, the dual space X^* is smooth of power type p^* (cf. Theorem 2.43). Therefore, together with Lemma 2.48 we have

$$D_{j_{p^*}}(x_n^*, x_{n+1}^*) \le 2^{p^*-1} \frac{G_{p^*}}{p^*}[\mu^{p^*}\|\psi_n\|^{p^*} + \beta_n^{p^*}\|x_n^*\|^{p^*}].$$

Together with Lemma 5.4 we have

$$
\begin{aligned}
&D_{j_p}(x_{n+1}, x^\dagger)\\
&\le (1-\beta_n)D_{j_p}(x_n, x^\dagger) - \mu\|Ax_n - y^\delta\|^p + \mu\delta\|Ax_n - y^\delta\|^{p-1}\\
&\quad + S\|Ax_n - y^\delta\|\beta_n + S\delta\beta_n + 2^{p^*-1}\frac{G_{p^*}}{p^*}[\mu^{p^*}\|\psi_n\|^{p^*} + \beta_n^{p^*}\|x_n^*\|^{p^*}].
\end{aligned}
$$

Assume that the iteration did not stop in last step, therefore the *stopping criterion* is not fulfilled for x_n. Hence, we have

$$\delta^{p-1} < \tau\beta_n.$$

Due to Young's inequality we have

$$
\begin{aligned}
\mu\delta\|Ax_n - y^\delta\|^{p-1} &\le \mu\|Ax_n - y^\delta\|^{p-1} \cdot \tau^{p^*-1}\beta_n^{p^*-1}\\
&\le \tfrac{1}{p^*}\mu^{p^*}\|Ax_n - y^\delta\|^p + \tfrac{1}{p}\tau^{p^*}\beta_n^{p^*}
\end{aligned}
$$

and

$$\|Ax_n - y^\delta\| \cdot \beta_n S \le \tfrac{1}{p}\mu^p\|Ax_n - y^\delta\|^p + \tfrac{1}{p^*}\beta_n^{p^*}\frac{S^{p^*}}{\mu^{p^*}}.$$

Since $\psi_n = A^*j_p(Ax_n - y^\delta)$ we get due to the properties of the duality mapping

$$2^{p^*-1}\frac{G_{p^*}}{p^*}\|\psi_n\|^{p^*} \le 2^{p^*-1}\frac{G_{p^*}}{p^*}\|A\|^{p^*}\|Ax_n - y^\delta\|^p.$$

Altogether we arrive at

$$
\begin{aligned}
&D_{j_p}(x_{n+1}, x^\dagger)\\
&\le (1-\beta_n)D_{j_p}(x_n, x^\dagger)\\
&\quad + \|Ax_n - y^\delta\|^p\left(-\mu + \tfrac{1}{p^*}\mu^{p^*} + \tfrac{1}{p}\mu^p + 2^{p^*-1}\frac{G_{p^*}}{p^*}\mu^{p^*}\|A\|^{p^*}\right)\\
&\quad + \tfrac{1}{p}\tau^{p^*}\cdot\beta_n^{p^*} + \tfrac{1}{p^*}\frac{S^{p^*}}{\mu^{p^*}}\cdot\beta_n^{p^*} + S\tau^{p^*-1}\cdot\beta_n^{p^*} + 2^{p^*-1}\frac{G_{p^*}}{p^*}\|x_n^*\|^{p^*}\cdot\beta_n^{p^*}.
\end{aligned}
$$

The bracket in the above inequality is negative by the initial assumption on μ, therefore the last estimation simplifies to

$$
\begin{aligned}
D_{j_p}&(x_{n+1}, x^\dagger) \\
&\leq (1 - \beta_n) D_{j_p}(x_n, x^\dagger) \\
&\quad + \left(\tfrac{1}{p} \tau^{p^*} + \tfrac{1}{p^*} \tfrac{S^{p^*}}{\mu^{p^*}} + S\tau^{p^*-1} + 2^{p^*-1} \tfrac{G_{p^*}}{p^*} \|x_n^*\|^{p^*} \right) \cdot \beta_n^{p^*} \\
&= (1 - \beta_n) D_{j_p}(x_n, x^\dagger) + \tfrac{\gamma_n}{p^*} \beta_n^{p^*}.
\end{aligned}
$$

What we have proven may be seen as part of an induction to show that $D_{j_p}(x_n, x^\dagger) \leq R_n$ for all n. We have by assumption $D_{j_p}(x_0, x^\dagger) \leq R_0$ and assuming $D_{j_p}(x_n, x^\dagger) \leq R_n$ we have proven that $D_{j_p}(x_{n+1}, x^\dagger) \leq R_{n+1}$. We notice that the number β_n is chosen such that the right-hand side of the last inequality is minimized. Therefore, we have $R_{n+1} \leq R_n$, the sequence (R_n) is monotonically decreasing and bounded, therefore convergent. Next, by Dunn's trick, cf. Theorem 4.18, we show that the sequence (R_n) is a zero sequence.

Since the sequence $(D_{j_q}(x_n, x^\dagger))$ is bounded, so are (x_n), cf. Theorem 2.48. Therefore, there exist two constants γ, Γ such that for all n we have

$$
0 < \gamma \leq \gamma_n \leq \Gamma < \infty.
$$

Therefore, we have

$$
\frac{1}{R_{n+1}^{p-1}} - \frac{1}{R_n^{p-1}} \geq \mathcal{C} \left(\beta_n - \tfrac{1}{p^*} \tfrac{\gamma_n}{R_n} \beta_n^{p^*} \right) \frac{1}{R_n^{p-1}} \geq \mathcal{C} \min\{ R_0^{-(p-1)}, \Gamma^{-(p-1)} \} \geq \mathcal{C} > 0
$$

and

$$
\frac{1}{R_n^{p-1}} \geq \sum_{k=0}^{n-1} \frac{1}{R_{k+1}^{p-1}} - \frac{1}{R_k^{p-1}} \geq \mathcal{C} \cdot n
$$

and finally

$$
R_n \leq \mathcal{C} \cdot n^{-\frac{1}{p-1}},
$$

where $\mathcal{C} > 0$ does not depend on n nor δ.

Therefore, we get

$$
\begin{aligned}
\beta_n &= \min\{1, \gamma_n^{-(p-1)} \cdot R_n^{p-1}\} \\
&\leq \min\{1, \gamma^{-(p-1)} \cdot n^{-1}\} \leq \mathcal{C} n^{-1}.
\end{aligned}
$$

Hence, the sequence (β_n) is a zero sequence. Therefore, for some sufficiently large n we have $\tau \beta_n \leq \delta^{p-1}$. This shows that $N(\delta, y^\delta)$ is well-defined.

Let $N(\delta, y^\delta)$ be the index for which the iteration terminates. Further, let $\delta^{p-1} < \frac{1}{\tau}$. Then, we have $\tau\beta_{N(\delta,y^\delta)} \leq \delta^{p-1} < \frac{1}{\tau}$. Hence, we get $\beta_{N(\delta,y^\delta)} < 1$. Therefore, we have

$$\beta_{N(\delta,y^\delta)} = \min\{1, (R_{N(\delta,y^\delta)}/\gamma_{N(\delta,y^\delta)})^{p-1}\} = (R_{N(\delta,y^\delta)}/\gamma_{N(\delta,y^\delta)})^{p-1}.$$

Finally, due to $\tau\beta_{N(\delta,y^\delta)} \leq \delta^{p-1}$ we have

$$D_{j_p}(x_{N(\delta,y^\delta)}, x^\dagger) \leq R_{N(\delta,y^\delta)} = \gamma_{N(\delta,y^\delta)} \cdot \beta_{N(\delta,y^\delta)}^{\frac{1}{p-1}} \leq C \cdot \delta.$$

And the proof is complete.   □

Remark 5.7. *The above modified Landweber iteration may be regarded as a blueprint for a plethora of other modified versions of (5.2). Some of these alternative versions of the above iteration are studied by the author and Hein in [32] and by Hein in [28].*

Further, we remark that the modified iteration may also be extended to the case of nonlinear operators, for more details cf. [32].

6

Numerical Experiments

The aim of this chapter is to present results of numerical experiments for the following algorithms:

1. Steepest descent in the primal space

2. Steepest descent in the dual space

3. Engl's discrepancy principle

4. (Modified) Landweber-iteration

6.1   Update in the primal space

We recall that the steepest descent in the primal space is given by

$$x_{n+1} := x_n - \mu_n J_{q^*}^{X^*}(\psi_n), \qquad \psi_n = \nabla \, \mathrm{T}_\alpha(x_n) = A^* J_p^Y(Ax_n - y^\delta) + \alpha J_q^X(x_n),$$

where the step size μ_n may be chosen either via a line search along the exact Tikhonov functional, i.e.

$$\mu_n := \mathrm{argmin}_\mu \, \mathrm{T}_\alpha(x_n - \mu J_{q^*}^{X^*}(\psi_n)) \tag{6.1}$$

or via a line search along a surrogate of the Tikhonov functional, namely

$$\mu_n := \mathrm{argmin}_\mu \{ -\mu \|\psi_n\|^{q^*} + |\mu|^p \tfrac{G_Y}{p} \|A J_{q^*}^{X^*}(\psi_n)\|^p + \alpha \cdot |\mu|^q \tfrac{G_X}{q} \|\psi_n\|^{q^*} \}. \tag{6.2}$$

In this section, we will call the line search of (6.1) *exact line search* and the line search of (6.2) *surrogate line search* and the related step size μ_n *exact step size* resp. *surrogate step size*.

Remark 6.1. *We remark that the exact line search*

$$\mu_n := \operatorname{argmin}_\mu \operatorname{T}_\alpha(x_n - \mu J_{q^*}^{X^*}(\psi_n))$$

may also be reformulated as the root finding problem

$$\langle \nabla \operatorname{T}_\alpha(x_n - \mu J_{q^*}^{X^*}(\psi_n)), -J_{q^*}^{X^*}(\psi_n)\rangle = 0.$$

The surrogate line search on the other hand

$$\mu_n := \operatorname{argmin}_\mu \{-\mu\|\psi_n\|^{q^*} + |\mu|^p \tfrac{G_Y}{p}\|AJ_{q^*}^{X^*}(\psi_n)\|^p + \alpha \cdot |\mu|^q \tfrac{G_X}{q}\|\psi_n\|^{q^*}\}.$$

may also be reformulated as the root finding problem

$$\operatorname{sign}(\mu) \cdot |\mu|^{p-1} G_Y \|AJ_{q^*}^{X^*}(\psi_n)\|^p + (-1 + \alpha \cdot \operatorname{sign}(\mu) \cdot |\mu|^{q-1} G_X)\|\psi_n\|^{q^*} = 0.$$

Example 6.2. *In the first step we want to compare the effects of choice of the step size on the iteration. To do so, we consider a simple two-dimensional example. Namely, we consider the linear, continuous operator $A : \mathbb{R}^2 \to \mathbb{R}^2$ given by*

$$Ax := \begin{pmatrix} 2 & 1 \\ 2 & 1 \end{pmatrix} x,$$

where the pre-image space is equipped with a ℓ^X-norm and the image space is equipped with the ℓ^2-norm. Then, the related Tikhonov functional is given by

$$\operatorname{T}_\alpha(x) = \tfrac{1}{2}\|Ax - y^\delta\|_2^2 + \alpha \tfrac{1}{q}\|x\|_X^q,$$

where

$$q = \min\{X, 2\}.$$

Further, we choose $y^\delta = (2, 2)^T$ and $(-1, 2.7)^T$ as the initial iterate x_0.

Remark 6.3. *In Figure 6.1, we show the results for exact line search and surrogate line search for $X = 1.1$ and $X = 11$ in Example 6.2. We observe that for $X = 1.1$ the algorithm with the exact step size converges much faster than the version with the surrogate step size. On the other hand, for $X = 11$ we see that both, the iteration with the exact and with the surrogate step size, converge at about the same rate.*

Let us now investigate the computational cost of the two line search strategies: First, we notice that in both cases the evaluation of the vector $AJ_{q^*}(\psi_n)$ is necessary.

The method with the exact step size converges considerably faster than the method with the surrogate step size. However, one should keep in mind that for every step of the exact line search one has to evaluate the norm of the pre-image space and the norm of the image space, which is not the case for the surrogate line search. Hence, this additional cost of evaluating the norms several times in every iteration

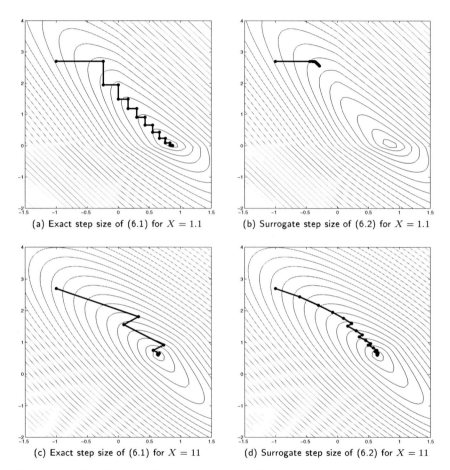

(a) Exact step size of (6.1) for $X = 1.1$

(b) Surrogate step size of (6.2) for $X = 1.1$

(c) Exact step size of (6.1) for $X = 11$

(d) Surrogate step size of (6.2) for $X = 11$

Figure 6.1: Visualization of the 50 first steps of the steepest descent in the primal space for the two-dimensional Example 6.2.

step only pays off if the evaluation of the operator A is much more expensive than the evaluation of the norms. We remind that this is the standing quiet assumption made throughout this thesis, cf. Remark 2.27.

Similar arguments hold for the formulation of the step sizes as solutions of root finding problems.

On the other hand, sometimes like for the (sequence) Besov spaces $B_{p,q}^s(\mathbb{R})$ resp. $b_{p,q}^s$ with $p \neq q$, we do not have any information on the the constants of smoothness G_X and G_Y. Then, we cannot use the surrogate line search and have to use the exact line search.

Remark 6.4. *The integral operator*

$$Af(x) = \int_0^x f(t)dt. \qquad f : [0, 1] \to \mathbb{R}$$

is a good benchmark for regularization algorithms in Banach spaces, cf. [59]. A discretization of this operator leads to the $(N \times N)$-matrix

$$A = \frac{1}{N} \begin{pmatrix} 1 & 0 & \dots & 0 \\ \vdots & \ddots & \ddots & \vdots \\ \vdots & & \ddots & 0 \\ 1 & \dots & \dots & 1 \end{pmatrix}.$$

Next, we want to compare convergence rates predicted by the theory with rates obtained numerically.

Example 6.5. *We consider here the (discretized) integral operator A, as it is defined in Remark 6.4. Since it is not possible to test the steepest descent method in the primal for every possible choice of norms on the pre-image and image space, we only consider the most prominent case where the image space is equipped with a (discretized) version of the $L^2(0, 1)$ norm and the pre-image space is equipped with a (discretized) version of the $L^X(0, 1)$ norm, where we use the same discretization for the norm as for the integral operator itself.*

We define the Tikhonov functional via

$$\mathrm{T}_\alpha(x) := \tfrac{1}{p}\|Ax - y^\delta\|_2^p + \alpha \cdot \tfrac{1}{q}\|x\|_X^q ,$$

with

$$p := 2 \qquad and \qquad q := \min\{2, X\}.$$

Therefore, the index p is the power of smoothness of the image space and the index q is the power of smoothness of the pre-image space. With this setting we may use the steepest descent in the primal space to minimize the Tikhonov functional.

For the analysis of convergence of the steepest descent we have to know the minimizer of the Tikhonov functional. However, in general this minimizer is not

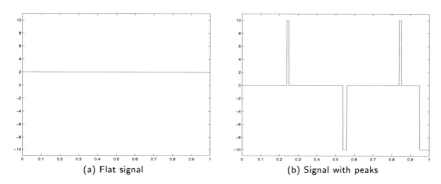

(a) Flat signal   (b) Signal with peaks

Figure 6.2: Signals used in Examples 6.5 and 6.9.

known.   We overcome this problem in the following way: First, we choose the minimizer and then generate the related noisy data from the optimality condition $A^* j_p(A x_\alpha^\delta - y^\delta) + \alpha j_q(x_\alpha^\delta) = 0$.   We use a flat signal and a signal with peaks, depicted in Figure 6.2, as minimizers to generate the noisy data y^δ.

The convergence results for resulting steepest descent in the primal space for $N = 1000$ and $\alpha = 10^{-3}$ are shown in Figure 6.3.   We observe that the results for $X < 2.0$ differ significantly from those for $X \geq 2.0$.

For $X < 2.0$ the iteration with the exact step size is always (in the long run) considerably faster than the iteration with the surrogate step size. However, surprisingly, for $X \geq 2.0$ the difference between the exact line search and the surrogate line search is negligible.   We recall that similar results were also true for the two-dimensional Example 6.2 (cf. Figure 6.1).   Therefore, we conclude that for $X < 2.0$ one should choose the exact step size due to the better (apparent) convergence behavior and for $X \geq 2.0$ one should choose the surrogate step size since in this case it is as good as the exact line search but computationally cheaper.

We observe that the apparent convergence rates obtained by the surrogate line search in our example match the convergence rates predicted by the theory.

6.2   Update in the dual space

We recall that the steepest descent in the dual space is given by

$$x_{n+1}^* := x_n^* - \mu_n \psi_n \text{ with } \psi_n \in \partial \mathrm{T}_\alpha(x_n) = A^* J_p^Y (A x_n - y^\delta) + \alpha J_q^X (x_n),$$
$$x_{n+1} := J_{q^*}^{X^*} (x_{n+1}^*),$$

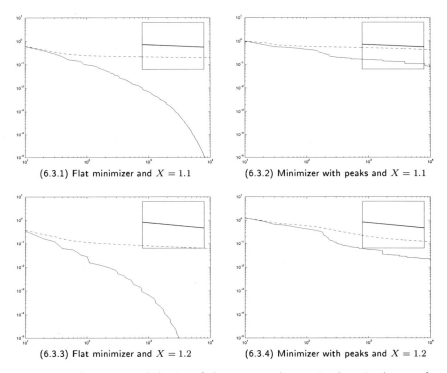

(6.3.1) Flat minimizer and $X = 1.1$ (6.3.2) Minimizer with peaks and $X = 1.1$

(6.3.3) Flat minimizer and $X = 1.2$ (6.3.4) Minimizer with peaks and $X = 1.2$

Figure 6.3: Convergence behavior of the steepest descent in the primal space for the Tikhonov functional of Example 6.5; for the exact step size (solid line) and the surrogate step size (dashed line); for both minimizers depicted in Figure 6.2. For $X \neq 2.0$, i.e. the sublinear cases, the solid line in the small subplot has the slope corresponding to the rate predicted by the theory. (x-axis: iteration index n; y-axis: norm $\|x_n - x_\alpha^\delta\|_X$)

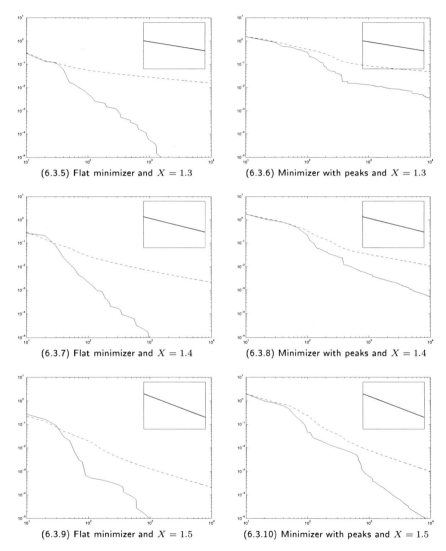

Figure 6.3: Convergence behavior of the steepest descent in the primal space (cont.)

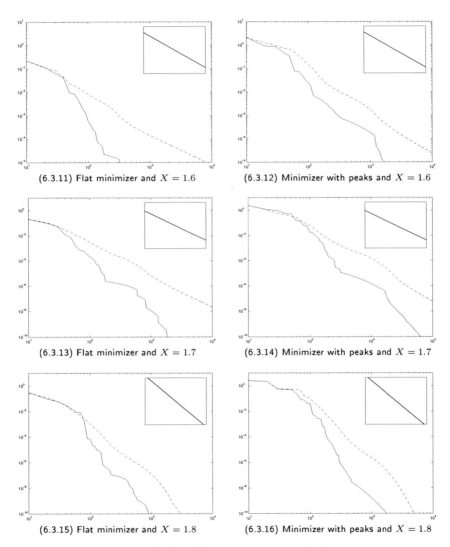

(6.3.11) Flat minimizer and $X = 1.6$ (6.3.12) Minimizer with peaks and $X = 1.6$

(6.3.13) Flat minimizer and $X = 1.7$ (6.3.14) Minimizer with peaks and $X = 1.7$

(6.3.15) Flat minimizer and $X = 1.8$ (6.3.16) Minimizer with peaks and $X = 1.8$

Figure 6.3: Convergence behavior of the steepest descent in the primal space (cont.)

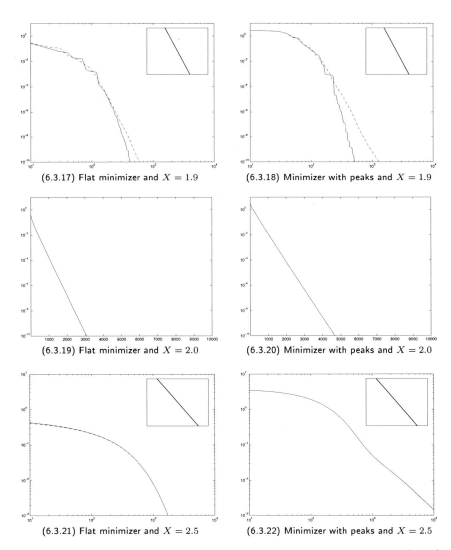

Figure 6.3: Convergence behavior of the steepest descent in the primal space (cont.)

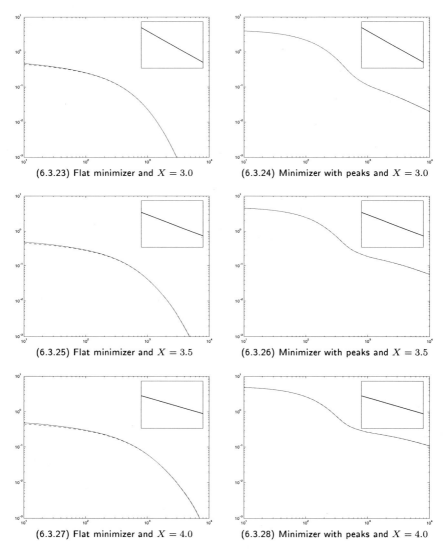

Figure 6.3: Convergence behavior of the steepest descent in the primal space (cont.)

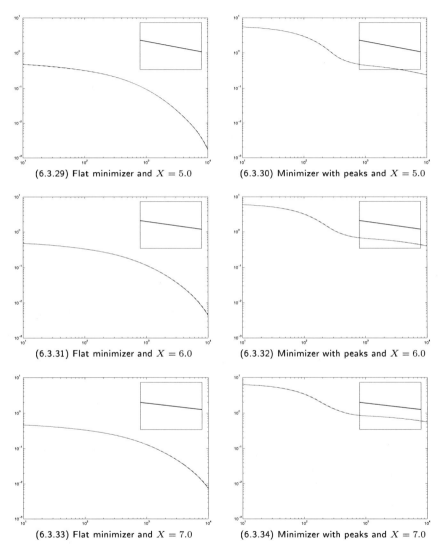

(6.3.29) Flat minimizer and $X = 5.0$

(6.3.30) Minimizer with peaks and $X = 5.0$

(6.3.31) Flat minimizer and $X = 6.0$

(6.3.32) Minimizer with peaks and $X = 6.0$

(6.3.33) Flat minimizer and $X = 7.0$

(6.3.34) Minimizer with peaks and $X = 7.0$

Figure 6.3: Convergence behavior of the steepest descent in the primal space (cont.)

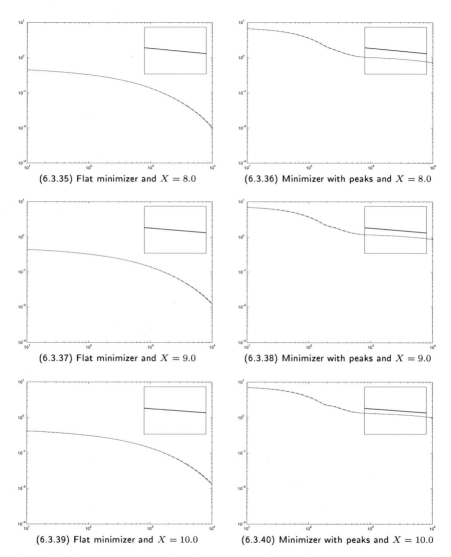

(6.3.35) Flat minimizer and $X = 8.0$ (6.3.36) Minimizer with peaks and $X = 8.0$

(6.3.37) Flat minimizer and $X = 9.0$ (6.3.38) Minimizer with peaks and $X = 9.0$

(6.3.39) Flat minimizer and $X = 10.0$ (6.3.40) Minimizer with peaks and $X = 10.0$

Figure 6.3: Convergence behavior of the steepest descent in the primal space (cont.)

where the step size μ_n may be chosen either via a line search along an estimation of the Bregman distance $D_{j_q}(x_{n+1}, x_\alpha^\delta)$

$$\mu_n := \operatorname{argmin}_\mu \{\max\{1 - \mu\alpha, 0\} R_n + D_{j_{q^*}}(x_n^*, x_n^* - \mu\psi_n)\} \qquad (6.3)$$

where the numbers R_n are updated by

$$R_{n+1} := (1 - \mu_n\alpha) R_n + D_{j_{q^*}}(x_n^*, x_n^* - \mu_n\psi_n),$$

or the step size μ_n my be chosen via a line search along another estimation of the Bregman distance $D_{j_q}(x_{n+1}, x_\alpha^\delta)$

$$\mu_n = \operatorname{argmin}_\mu \max\{1 - \mu\alpha, 0\} R_n + |\mu|^{q^*} \frac{G_{q^*}}{q^*} \cdot \|\psi_n\|^{q^*}$$

$$= \min\left\{ \left(\frac{\alpha}{G_{q^*}} \cdot \frac{R_n}{\|\psi_n\|^p} \right)^{\frac{1}{p-1}}, \frac{1}{\alpha} \right\} \qquad (6.4)$$

where the numbers R_n are updated via

$$R_{n+1} = (1 - \mu_n\alpha) R_n + \mu_n^{q^*} \frac{G_{q^*}}{q^*} \cdot \|\psi_n\|^{q^*}.$$

In analogy to the last section we will call the first step size *exact step size* and the second step size *surrogate step size;* we will call the related line searches *exact line search* resp. *surrogate line search.*

Remark 6.6. *We remark that the exact line search*

$$\operatorname{argmin}_\mu \{\min\{1 - \mu\alpha, 0\} R_n + D_{j_{q^*}}(x_n^*, x_n^* - \mu\psi_n)\}$$

may also be formulated as the root finding problem

$$\partial_\mu (\max\{1 - \mu\alpha, 0\} R_n) + \langle j_{q^*}(x_n^* - \mu\psi_n) - x_n, -\psi_n \rangle \ni 0.$$

The above is true since,

$$\partial_\mu D_{j_{q^*}}(x_n^*, x_n^* - \mu\psi_n) = \langle j_{q^*}(x_n^* - \mu\psi_n) - x_n, -\psi_n \rangle.$$

Further, the function $\mu \mapsto \max\{1 - \mu\alpha, 0\} R_n$ is convex and

$$\partial_\mu(\max\{1 - \mu\alpha, 0\}) = \begin{cases} -\alpha, & \mu < \frac{1}{\alpha}, \\ [-\alpha, 0], & \mu = \frac{1}{\alpha}, \\ 0, & \mu > \frac{1}{\alpha}. \end{cases}$$

Therefore, if $\langle j_{q^*}(x_n^* - \frac{1}{\alpha}\psi_n) - x_n, -\psi_n \rangle \in [0, \alpha R_n]$ we may set $\mu_n = \frac{1}{\alpha}$, else let μ_n be a positive solution of

$$-\alpha R_n + \langle j_{q^*}(x_n^* - \frac{1}{\alpha}\psi_n) - x_n, -\psi_n \rangle = 0.$$

Example 6.7. *As in the last section we start with a simple two dimensional example. We use the setting of Example 6.2, excerpt for the Tikhonov functional which shall be given this time by*

$$\mathrm{T}_\alpha(x) = \tfrac{1}{2}\|Ax - y^\delta\|_2^2 + \alpha \tfrac{1}{q}\|x\|_X^q \ ,$$

where this time we set

$$q = \max\{X, 2\}.$$

Remark 6.8. *In Figure 6.4, we show the results for the setting of Example 6.7 with the exact step size and surrogate step size for $X = 1.1$ and $X = 5$. First, we observe that the iteration process for the steepest descent in the dual space is much more wiggly than the steepest descent in the primal space (cf. Figure 6.1).*

For $X = 1.1$ we notice that the iteration with the exact step size and the iteration with the surrogate step size seem to converge at the same rate. For $X = 5$ the iteration with the exact step size converged almost to the minimizer of the function, whereas the iteration with the surrogate step size gets stuck at about $(-1.0, 0.8)^T$. Hence, the situation is opposite to steepest descent in the primal space, where both step sizes performed similarly for $X = 5$ and where, for $X = 1.1$, the exact step size converged considerably faster.

Next, we consider the computational cost of the two line search strategies: To get the step size by the exact line search we have to evaluate the Bregman distance $D_{j_{q^*}}(x_n^*, x_n^* - \mu_n \psi_n)$. Every evaluation of the Bregman distance implies evaluation of two norms, a duality mapping and a dual composition. On the other hand, to get the step size by the surrogate line search one has to know the constant G_{q^*} and evaluate the norm $\|\psi_n\|^{q^*}$. Therefore, the same considerations as for the steepest descent in the primal space apply also here: The additional cost of evaluating the norms, the duality mapping and the dual composition several times in every iteration step only pays off, if the evaluation of the operator is much more expensive than the evaluation of the norms (resp. dual composition). As before this is the standing quiet assumption made throughout this thesis, cf. Remark 2.27.

Similar considerations hold also for the root finding problem formulation of the line search.

If we do not know G_{q^*}, like for the (sequence) Besov spaces $B_{p,q}^s(\mathbb{R})$ resp. $b_{p,q}^s$ with $p \neq q$, we cannot use the surrogate line search and have to use the exact line search regardless of the additional cost.

Next, we shall compare the convergence rates predicted by the theory to the rates, which are obtained numerically with the discretized integral operator introduced in Remark 6.4:

Example 6.9. *As for the steepest descent in the primal space, we only consider the most prominent case, where the image space is equipped with a (discretized) version*

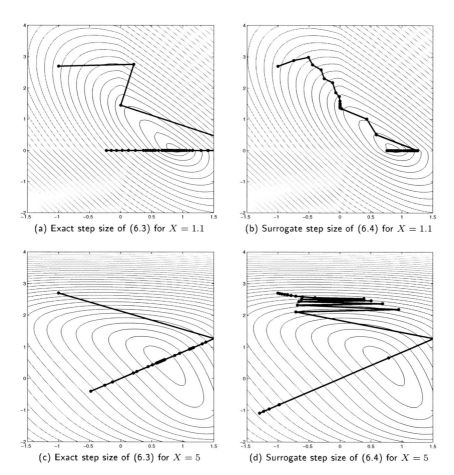

(a) Exact step size of (6.3) for $X = 1.1$ (b) Surrogate step size of (6.4) for $X = 1.1$

(c) Exact step size of (6.3) for $X = 5$ (d) Surrogate step size of (6.4) for $X = 5$

Figure 6.4: Visualization of the 50 first steps of the steepest descent in the dual space for the two-dimensional Example 6.7.

of the $L^2(0, 1)$ norm and the pre-image space is equipped with a (discretized) version of the $L^X(0, 1)$ norm, where we use the same discretization for the norm as for the integral operator itself.

The space $L^2(0, 1)$ is convex of power type 2. Further, the space $L^X(0, 1)$ is convex of power type $\max\{X, 2\}$. In order to apply the steepest descent in the dual to minimize the Tikhonov functional we define it via

$$\mathrm{T}_\alpha(x) := \tfrac{1}{p}\|Ax - y^\delta\|_2^p + \alpha \cdot \tfrac{1}{q}\|x\|_X^q \,,$$

with

$$p := 2 \qquad and \qquad q := \max\{2, X\}.$$

As in Example 6.5, we first choose the minimizer of the Tikhonov functional and then generate the related y^δ by the method described in Example 6.5. We again use the flat signal and the signal with peaks, cf. Figure 6.2, as minimizers. We test the steepest descent in the dual space for $N = 1000$ and $\alpha = 10^{-3}$.

In Figure 6.5a, we depict a typical behavior for the sequence $\|x_n - x_\alpha^\delta\|$. It is easy to see that the sequence is highly oscillating, therefore instead of the original sequence we will use the (monotonically descending, upper) envelope. For an (infinite) vector a_n the (monotonically descending, upper) envelope b_n may be defined via

$$b_n = \sup_{k \geq n} a_k.$$

The envelope of the norms of Figure 6.5a is shown in Figure 6.5b.

In Figure 6.6, we show the envelopes of the sequence of norms $\|x_n - x_\alpha^\delta\|$ for a range of values of X. It can be observed that the results for $X \leq 2.0$ are different from those for $X > 2.0$.

For $X \leq 2.0$ the algorithm converges linearly to the minimizer, which is in accordance with the theoretical prediction. Moreover, we observe that both, the method with the exact step size and the method with the surrogate line search converge at a similar rate. In some cases the exact step size performs better than the surrogate step size, whereas in other cases the opposite is true. Hence, we conclude that there is no a-priori preference about the choice of the step size for $X \leq 2.0$. However, in the light of computational cost, we may say that for $X \leq 2.0$ the surrogate line search should be preferred to the exact line search.

The situation is different for $X > 2.0$. Here, the exact line search always performs better in the long run than the surrogate line search. Therefore, for $X > 2.0$ the exact line search should be considered as superior to the surrogate line search, as long as the constraints on the computational cost of the evaluation of the operator, discussed in Remark 6.8, hold. Further, we also observe that due to the high oscillatory behavior of the sequences $\|x_n - x_\alpha^\delta\|$ its very hard to compare for $X > 2.0$ the convergence rates obtained by the iteration in numerical experiments with the convergence rates predicted by the theory .

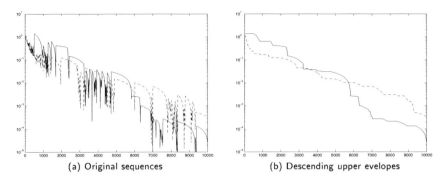

(a) Original sequences   (b) Descending upper evelopes

Figure 6.5: Two sequences and their descending upper envelopes.

Remark 6.10. *In the final part of this section, we shall discuss two questions which arise in the application of the steepest descent in the dual space.*

In the initial step of the algorithm one has to chose the number R_0 such that $D_{j_q}(x_0, x_\alpha^\delta) \leq R_0$. However, in applications the value $D_{j_q}(x_0, x_\alpha^\delta)$ is not known, even worse, in most applications even a sensible order of magnitude for $D_{j_q}(x_0, x_\alpha^\delta)$ is not known. Therefore, one always tends to overestimate the number $D_{j_q}(x_0, x_\alpha^\delta)$, probably by several orders of magnitude. Therefore, the most apparent question is: How does this overestimation affect the convergence of the algorithm? We show the effects of overestimating $D_{j_q}(x_0, x_\alpha^\delta)$ in the integral example (cf. Example 6.9) with $X = 1.1$. We choose $R_0 = D_{j_q}(x_0, x_\alpha^\delta)$, $R_0 = 10 \cdot D_{j_q}(x_0, x_\alpha^\delta)$, $R_0 = 1\,000 \cdot D_{j_q}(x_0, x_\alpha^\delta)$, and $R_0 = 1\,000\,000 \cdot D_{j_q}(x_0, x_\alpha^\delta)$. In Figure 6.7 the convergence results for the numbers R_n and the norms $\|x_n - x_\alpha^\delta\|$ for both the exact and the surrogate line search are presented. For all four choices of R_0 we conclude that, in the long run, the effects of the initial choice of R_0 on the convergence are negligible, even if R_0 is heavily overestimated.

We also ask: When is it most advisable to stop the iteration? As we can see in Figure 6.8 the graph of the numbers R_n has a stair-like structure, i.e. the numbers change only slightly for several iterations just to make a huge descent at a single step. Moreover, it can be observed that every stair-like step has an effect on the norm $\|x_n - x_\alpha^\delta\|$. In the zoomed version we can see that at every step "down the stairs" the norm increases. Therefore, we conclude: The iteration should only be stopped after a step "down the stairs" and one should take the iteration before the last one as the final iterate.

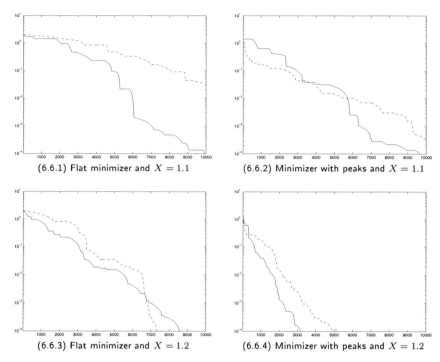

(6.6.1) Flat minimizer and $X = 1.1$ (6.6.2) Minimizer with peaks and $X = 1.1$

(6.6.3) Flat minimizer and $X = 1.2$ (6.6.4) Minimizer with peaks and $X = 1.2$

Figure 6.6: Convergence behavior for the steepest descent in the dual space for the Tikhonov functional of Example 6.9; for the exact step size (solid line) and the surrogate step size (dashed line); for both minimizers depicted in Figure 6.2. For $X > 2.0$, i.e. the sublinear cases, the solid line in the small subplot has the slope corresponding to the rate predicted by the theory. (x-axis: iteration index n; y-axis: norm $\|x_n - x_\alpha^\delta\|_X$)

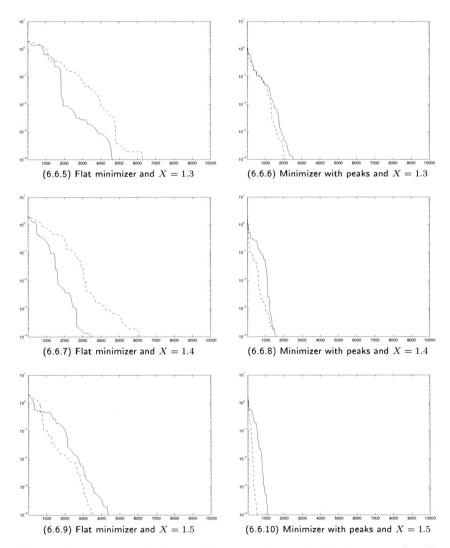

(6.6.5) Flat minimizer and $X = 1.3$

(6.6.6) Minimizer with peaks and $X = 1.3$

(6.6.7) Flat minimizer and $X = 1.4$

(6.6.8) Minimizer with peaks and $X = 1.4$

(6.6.9) Flat minimizer and $X = 1.5$

(6.6.10) Minimizer with peaks and $X = 1.5$

Figure 6.6: Convergence behavior of the steepest descent in the dual space (cont.)

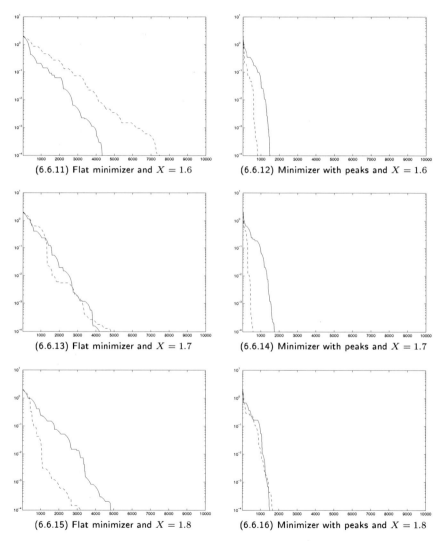

(6.6.11) Flat minimizer and $X = 1.6$ (6.6.12) Minimizer with peaks and $X = 1.6$

(6.6.13) Flat minimizer and $X = 1.7$ (6.6.14) Minimizer with peaks and $X = 1.7$

(6.6.15) Flat minimizer and $X = 1.8$ (6.6.16) Minimizer with peaks and $X = 1.8$

Figure 6.6: Convergence behavior of the steepest descent in the dual space (cont.)

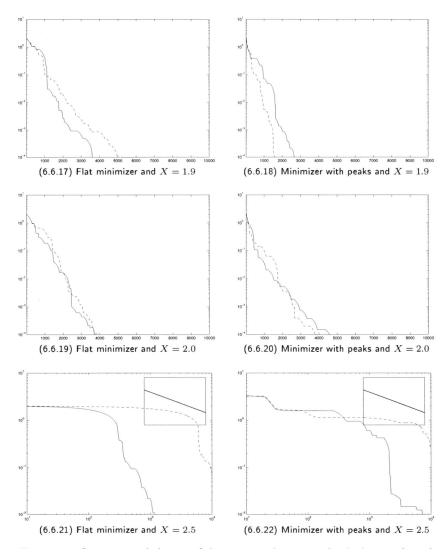

(6.6.17) Flat minimizer and $X = 1.9$

(6.6.18) Minimizer with peaks and $X = 1.9$

(6.6.19) Flat minimizer and $X = 2.0$

(6.6.20) Minimizer with peaks and $X = 2.0$

(6.6.21) Flat minimizer and $X = 2.5$

(6.6.22) Minimizer with peaks and $X = 2.5$

Figure 6.6: Convergence behavior of the steepest descent in the dual space (cont.)

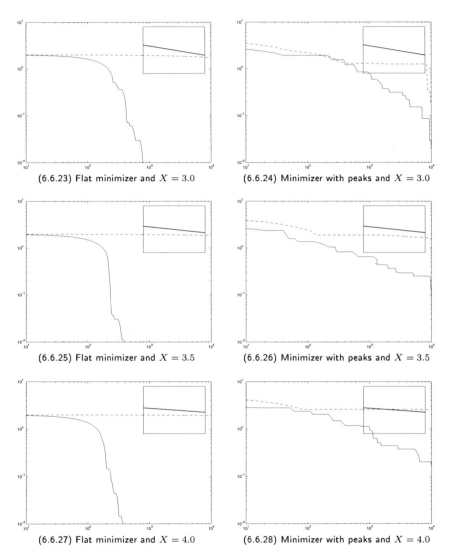

Figure 6.6: Convergence behavior of the steepest descent in the dual space (cont.)

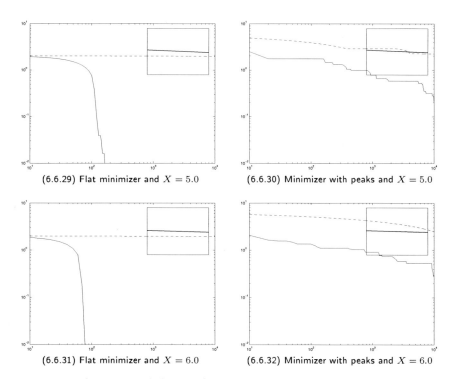

Figure 6.6: Convergence behavior of the steepest descent in the dual space (cont.)

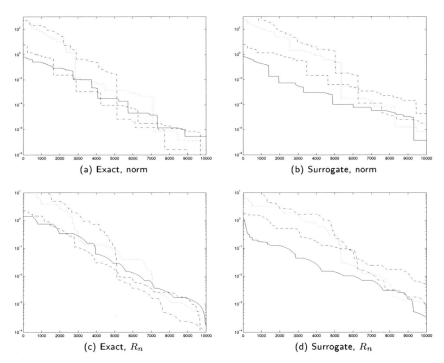

(a) Exact, norm   (b) Surrogate, norm

(c) Exact, R_n (d) Surrogate, R_n

Figure 6.7: Effects of the choice of R_0 on the convergence in norm and in R_n for exact line search and surrogate line search for $R_0 = D_{j_q}(x_0, x_\alpha^\delta)$ (solid line), $R_0 = 10 \cdot D_{j_q}(x_0, x_\alpha^\delta)$ (dashed line), $R_0 = 1\,000 \cdot D_{j_q}(x_0, x_\alpha^\delta)$ (dotted line) and $R_0 = 1\,000\,000 \cdot D_{j_q}(x_0, x_\alpha^\delta)$ (dashdot line). (x-axis: iteration index n; y-axis: norm $\|x_n - x_\alpha^\delta\|_X$ resp. number R_n)

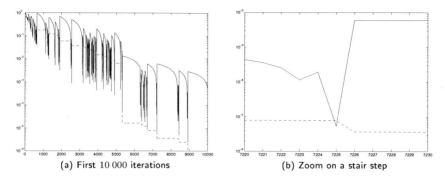

(a) First 10 000 iterations   (b) Zoom on a stair step

Figure 6.8: Visualization of the stair-like structure in the graph of the R_n (dashed line) together with the norm $\|x_n - x_\alpha^\delta\|$ (solid line) for the first 10 000 iterations and a zoom on a stair-like step. (x-axis: iteration index n; left y-axis: norm $\|x_n - x_\alpha^\delta\|_X$; right y-axis: number R_n)

6.3 Engl's discrepancy principle

We recall, cf. Section 4.3.3, that Engl's discrepancy principle consists of finding $\alpha(\delta, y^\delta)$ such that

$$\|A^* j_p(A x_{\alpha(\delta, y^\delta)}^\delta - y^\delta)\|^{q^*} = \delta^r \alpha(\delta, y^\delta)^{-s}$$

for appropriately chosen parameters p, q, r, s. In this section, we introduce a new method for solving the above problem and compare the results predicted by the theory with numerical results.

The new method for solving the problem of Engl's discrepancy principle is based on the so-called model functions, cf. [69] and the references therein, which were developed for solving the problem in the discrepancy principle of Morozov. First, we will give a short introduction to model functions and then present our algorithm.

Remark 6.11 (Model functions). *We consider the Tikhonov functional*

$$T_\alpha(x) = \tfrac{1}{p} \|Ax - y^\delta\|^p + \alpha \tfrac{1}{q} \|x\|^q$$

and the function

$$F(\alpha) := \min_x T_\alpha(x) = T_\alpha(x_\alpha^\delta).$$

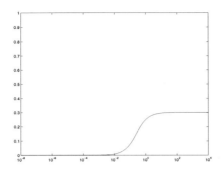

Figure 6.9: Typical graph of the function $t(\alpha)$, cf. Remark 6.11.

Formally differentiating the function F with respect to α we obtain

$$
\begin{aligned}
F'(\alpha) &= \frac{\partial}{\partial \alpha}(\frac{1}{p}\|Ax_\alpha^\delta - y^\delta\|^p) + \alpha\frac{\partial}{\partial \alpha}(\frac{1}{q}\|x_\alpha^\delta\|^q) + \frac{1}{q}\|x_\alpha^\delta\|^q \\
&= \langle\frac{\partial}{\partial x_\alpha^\delta}(\frac{1}{p}\|Ax_\alpha^\delta - y^\delta\|^p), \frac{\partial}{\partial \alpha}x_\alpha^\delta\rangle + \alpha\langle\frac{\partial}{\partial x_\alpha^\delta}(\frac{1}{q}\|x_\alpha^\delta\|^q), \frac{\partial}{\partial \alpha}x_\alpha^\delta\rangle + \frac{1}{q}\|x_\alpha^\delta\|^q \\
&= \langle\frac{\partial}{\partial x_\alpha^\delta}(\mathrm{T}_\alpha(x_\alpha^\delta)), \frac{\partial}{\partial \alpha}x_\alpha^\delta\rangle + \frac{1}{q}\|x_\alpha^\delta\|^q.
\end{aligned}
$$

Since x_α^δ is the minimizer of T_α we have $\frac{\partial}{\partial x_\alpha^\delta}(\mathrm{T}_\alpha(x_\alpha^\delta)) = 0$. Therefore,

$$
F'(\alpha) = \frac{1}{q}\|x_\alpha^\delta\|^q.
$$

If X and Y are Hilbert spaces and $p = q = 2$ then

$$
F(\alpha) + \alpha F'(\alpha) + \frac{1}{2}\|Ax_\alpha^\delta\|^2 = \frac{1}{2}\|y^\delta\|^2.
$$

Let the auxiliary function $t(\alpha)$ be defined as

$$
\frac{1}{2}\|Ax_\alpha^\delta\|^2 = t(\alpha) \cdot \frac{1}{2}\|x_\alpha^\delta\|^2 = t(\alpha) \cdot F'(\alpha).
$$

Then, the above equation can be rewritten in the following form

$$
F(\alpha) + (\alpha + t(\alpha)) \cdot F'(\alpha) = \frac{1}{2}\|y^\delta\|^2.
$$

The main assumption of the model functions approach is that the function t is approximately constant in the neighborhood of the solution of the discrepancy principle. As we can see in Figure 6.9 this assumption is not unreasonable. Hence, we replace the function $t(\alpha)$ by a constant t then the resulting differential equation is given by

$$
F(\alpha) + (\alpha + t) \cdot F'(\alpha) \approx \frac{1}{2}\|y^\delta\|^2.
$$

The equation

$$m(\alpha) + (\alpha + t) \cdot m'(\alpha) = \tfrac{1}{2}\|y^\delta\|^2.$$

has the solution

$$m(\alpha) = \tfrac{1}{2}\|y^\delta\|^2 + \frac{c}{\alpha + t}.$$

The functions of above type are called model functions, *since they are models of the function* F.

Remark 6.12 (Model functions in Engl's discrepancy principle). *Here, we explain how model functions may be utilized in order to to solve the problem of Engl's discrepancy principle, which aims to find* $\alpha(\delta, y^\delta)$ *such that*

$$\|A^* j_p(Ax^\delta_{\alpha(\delta,y^\delta)} - y^\delta)\|^{q^*} = \delta^r \alpha(\delta, y^\delta)^{-s}.$$

The main key to this is to recall that

$$\|A^* j_p(Ax^\delta_{\alpha(\delta,y^\delta)} - y^\delta)\|^{q^*} = \alpha^{q^*} \cdot q \cdot \tfrac{1}{q}\|x^\delta_{\alpha(\delta,y^\delta)}\|^q.$$

With $F(\alpha) := T_\alpha(x^\delta_\alpha)$ *we have* $F'(\alpha) = \tfrac{1}{q}\|x^\delta_\alpha\|^q$. *Hence, Engl's discrepancy principle may be reformulated as*

$$q \cdot \alpha(\delta, y^\delta)^{q^*+s} F'(\alpha(\delta, y^\delta)) = \delta^r.$$

We define the model functions

$$m_k(\alpha) = \tfrac{1}{p}\|y^\delta\|^p + \frac{c_k}{\alpha + t_k},$$

where c_k *and* t_k *are chosen such that*

$$m_k(\alpha_k) \quad = \quad F(\alpha_k) = T_{\alpha_k}(x^\delta_{\alpha_k})$$

and

$$m'_k(\alpha_k) \quad = \quad F'(\alpha_k) = \tfrac{1}{q}\|x^\delta_{\alpha_k}\|^q.$$

Then, starting with some guess α_k *one may define* α_{k+1} *as the solution of*

$$q \cdot \alpha^{q^*+s} m'_k(\alpha) = \delta^r.$$

Since $m'(\alpha) = \frac{-c_k}{(\alpha+t_k)^2}$, *the above equation is equivalent to*

$$\delta^r(\alpha + t_k)^2 + c_k \cdot q \cdot \alpha^{q^*+s} = 0$$

for $\alpha \neq t_k$. *Notice that since* $m'(\alpha_k) = \tfrac{1}{q}\|x^\delta_\alpha\|^q > 0$, *we have that* $c_k < 0$. *Therefore, the above equation has a solution if*

$$2 < q^* + s. \tag{6.5}$$

In our numerical experiments we have chosen

$$s = 1.$$

Then, the value of r is given by $r = (s+q^*) \cdot (p-1)$ *for the low order source condition* $j_q(x^\dagger) \in A^*\omega$, *resp.* $r = (s + q^*) \cdot p(p - 1)/(p + q - 1)$ *for the high order source condition* $j_q(x^\dagger) \in A^*j_p(A\omega)$ *(cf. Theorem 4.16). We justify this choice as follows: If the pre-image and the image space are Hilbert spaces and the source condition* $x^\dagger = A^*A\omega$ *holds and we choose* $q = p = 2$ *we get* $r = 2$, *which is the original choice of parameters* r *and* s *in [22]. Further, with this choice of* s *the condition (6.5) always holds, since* q *is chosen as the power of convexity of the pre-image space, hence* $q^* > 1$.

Finally, we remark that the proof of convergence for the above iterative scheme is a subject of ongoing research. However, the results of [37] obtained with model functions for the Morozov's discrepancy principle, allow us to expect that similar results may be obtained for the model functions approach for the discrepancy principle of Engl.

Example 6.13. *We use again the integral operator of Remark 6.4 with image space* $L^2(0,1)$ *and pre-image space* $L^X(0,1)$. *We only test the convergence rates under high order source condition, i.e. the minimum norm solution is chosen such that the source condition* $j_q(x^\dagger) = A^*j_p(A\omega)$ *holds, where* $p = 2$ *is the degree of smoothness of* Y *and* $q = \min\{2, X\}$ *is the degree of smoothness of* X. *Then the related Tikhonov functional is given by*

$$\mathrm{T}_\alpha(x) := \tfrac{1}{p}\|Ax - y^\delta\|_Y^2 + \alpha \cdot \tfrac{1}{q}\|x\|_X^2 \qquad \text{with } p = 2, q = \min\{2, X\}.$$

Further, we choose $s = 1$ *and* r *such that*

$$\frac{r}{s + q^*} = \frac{p}{p + q - 1} \cdot (p - 1).$$

Moreover, we use the general form of the Engl's discrepancy principle, i.e. we choose $\alpha(\delta, y^\delta)$ *such that*

$$\|A^*j_p(Ax^\delta_{\alpha(\delta,y^\delta)} - y^\delta)\|^{q^*} = C \cdot \delta^r \alpha(\delta, y^\delta)^{-s},$$

with $C > 0$, *cf. Remark 4.12. In particular, we chose the constant* C *in the range of* 10^3-10^4, *depending on the particular value of* X.

In Figure 6.10 the resulting convergence rates are depicted. It is easy to observe that the resulting convergence rates are at least as good as the rates predicted by the theory.

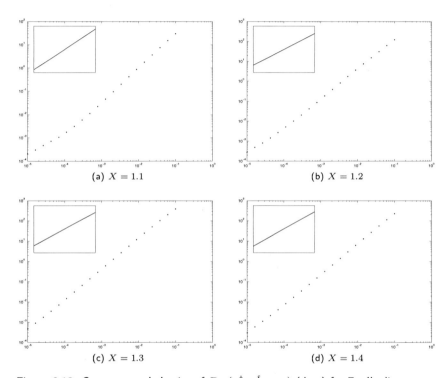

Figure 6.10: Convergence behavior of $D_{j_q}(x^\dagger, x^\delta_{\alpha(\delta,y^\delta)})$ (dots) for Engl's discrepancy principle under high order source condition and for different pre-image spaces. The solid line in the small subplots has the slope corresponding to the rate predicted by the theory. (x-axis: noise level δ; y-axis: Bregman distance $D_{j_q}(x^\dagger, x^\delta_{\alpha(\delta,y^\delta)})$)

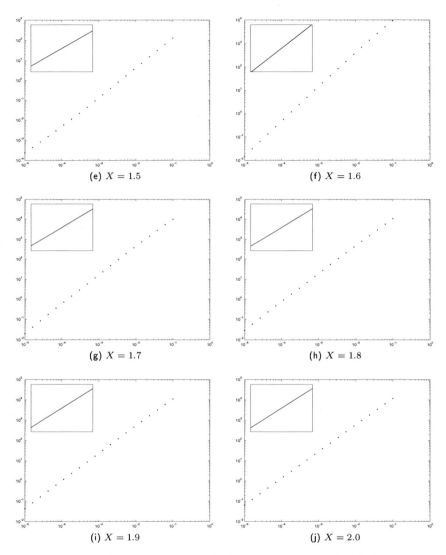

Figure 6.10: Convergence results for Engl's discrepancy principle (cont.)

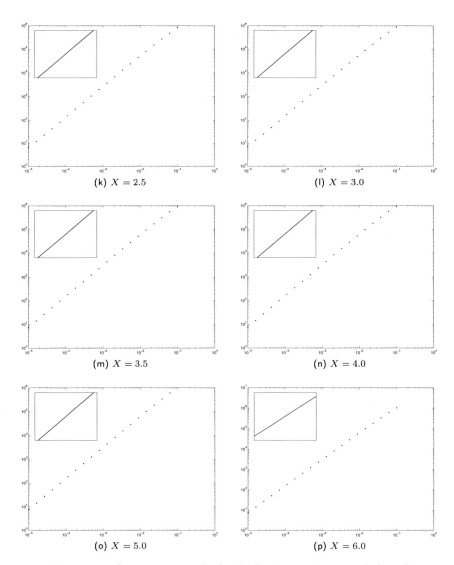

(k) $X = 2.5$

(l) $X = 3.0$

(m) $X = 3.5$

(n) $X = 4.0$

(o) $X = 5.0$

(p) $X = 6.0$

Figure 6.10: Convergence results for Engl's discrepancy principle (cont.)

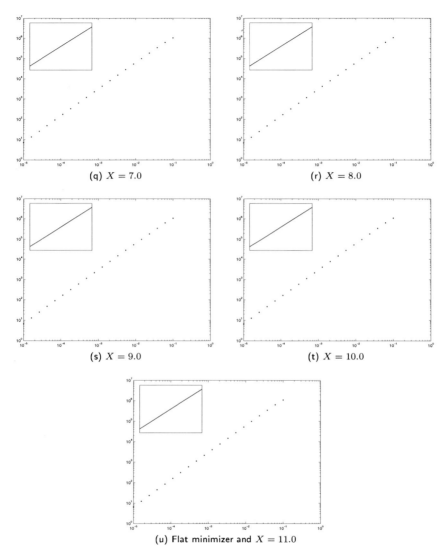

Figure 6.10: Convergence results for Engl's discrepancy principle (cont.)

6.4   (Modified) Landweber iteration

We recall that the the Landweber iteration in Banach spaces is given by

$$x_{n+1}^* := x_n^* - \mu_n \psi_n \qquad \psi_n := A^* j_p^Y (Ax_n - y^\delta) \in \partial(\tfrac{1}{p}\|A \cdot -y^\delta\|^p)(x_n)$$
$$x_{n+1} := J_{q^*}^{X^*}(x_{n+1}^*),$$
(6.6)

where the step size μ_n may be chosen via the line search

$$\mu_n^* = \operatorname{argmin}_\mu \{-\mu\|Ax_n - y^{\delta_k}\|^p + |\mu|\delta\|Ax_n - y^{\delta_k}\|^{p-1} + D_{j_{q^*}}(x_n^*, x_n^* - \mu\psi_n)\}$$
$$\mu_n = \min\{\mu_n^*, \overline{\mu}\|Ax_n - y^{\delta_k}\|^{q-p}\},$$

or via

$$\mu_n = \operatorname{argmin}_\mu \{-\mu_n\|Ax_n - y^\delta\|^p + |\mu_n|\delta\|Ax_n - y^\delta\|^{p-1} + |\mu_n|^{q^*}\tfrac{G_{q^*}}{q^*}\|\psi_n\|^{q^*}\}$$
$$= \left[(\|Ax_n - y^\delta\|^p - \delta\|Ax_n - y^\delta\|^{p-1})/(G_{q^*}\|\psi_n\|^{q^*})\right]^{q-1}$$

and the iteration is stopped whenever the iterate x_n fulfills $\|Ax_n - y^\delta\| \le \tau\delta$ with $\tau > 1$. We will call the first step size *exact step size* and the second step size *surrogate step size* and the related line searches *exact line search* resp. *surrogate line search*.

Remark 6.14. *We notice that due to*

$$\partial_\mu D_{j_{q^*}}(x_n^*, x_n^* - \mu\psi_n) = \langle j_{q^*}(x_n^* - \mu\psi_n) - x_n, -\psi_n\rangle$$

the exact line search may also be reformulated as the problem of finding a positive root of the equation

$$\|Ax_n - y^\delta\|^p + \delta\|Ax_n - y^\delta\|^{p-1} + \langle j_{q^*}(x_n^* - \mu\psi_n) - x_n, -\psi_n\rangle = 0.$$

Remark 6.15. *The Landweber iteration in Banach spaces was, to the best of author's knowledge, introduced in [59]. The authors provide results for the case, where the pre-image space is the space $L^q(0,1)$ and the step size is given by*

$$\mu_0 = C(1 - \tfrac{1}{\tau})^{q-1}\frac{(q^*)^{q-1}}{\|A\|^q}\|Ax_0 - y^\delta\|^{q-p}$$

if $x_0 = 0$ and

$$\mu_n = \frac{\tau_n}{\|A\|} \cdot \frac{\|x_n\|^{q-1}}{\|Ax_n - y^\delta\|^{p-1}}$$

else, where

$$\tau_n = \min\{1, \left(\frac{C(1 - \tfrac{1}{\tau})}{2^{q^*}G_{q^*}\|A\|} \cdot \frac{\|Ax_n - y^\delta\|}{\|x_n\|}\right)^{\frac{1}{\min(q^*,2)-1}}\}$$

| | Original | | | Adapted | |
	$C = 0.5$	$C = .99$	$C = 4.0$	surrogate	exact
$\tau = 10$	1079	535	133	48	49
$\tau = 4$	4336	2162	498	109	109
$\tau = 2$	10857	5396	1300	190	195
$\tau = 1.5$	19926	9946	2422	287	249
$\tau = 1.05$	168422	86298	20957	487	558
$\tau = 1.001$	8359602	4151363	1046979	952	1136

Table 6.1: Stopping indices $N(\delta, y^\delta)$ for different step size choices in the Landweber iteration.

with $G_{q^*} = \max\{2^{2-q^*}, q^* - 1\}$, $C < 1$.

Notice that the main difference between the algorithm considered in this thesis and the one described in [59] is the meaning of the number q. In the algorithm (6.6) the number q is the power of convexity of the pre-image space, whereas in - the numerical results section of - [59] the number q is the index of the sequence space $L^q(0,1)$. Fortunately, the authors of [59] focus on the case where the pre-image space is $L^2(0,1)$, for which both definitions for q coincide, since $L^2(0,1)$ is convex of power type 2. Therefore, we can compare the number of iterations needed in the original iteration to the number of iterations needed in our algorithm.

The image space is equipped with the norm $L^\infty(0,1)$ and $p = 2$. The authors choose A to be the (discretized) integral operator, see Remark 6.4.

In Table 6.1 the stopping indices for different choices of step sizes in the Landweber iteration are presented. It can be seen that the algorithm developed in this thesis converges faster than the one introduced in [59].

Remark 6.16. We conclude the chapter on numerical examples with a remark on the modified Landweber iteration. We use again the integral example of Remark 6.4. Here, we use the pre-image space $L^{1.5}(0,1)$ and the image space $L^2(0,1)$. As we can see in Figure 6.11 the rate predicted by theory is achieved in our case if the source condition $j_q(x^\dagger) = A^*\omega$ holds.

However, as we can see, cf. Figure 6.11, the value of stopping indices $N(\delta, y^\delta)$ explodes for small values of the noise-level δ. Therefore, we give here only a proof of concept for the modified Landweber iteration. In our opinion the next step in the analysis of the modified Landweber iteration must be further acceleration. Only then an extensive numerical study will give satisfactory outcome. For further numerical results concerning the modified Landweber iteration we refer to [32] and [28].

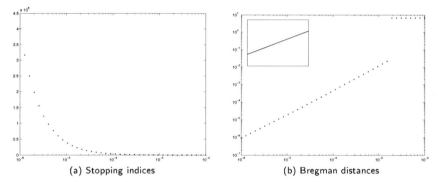

(a) Stopping indices   (b) Bregman distances

Figure 6.11: Modified Landweber iteration: Stopping indices $N(\delta, y^\delta)$ (dots, left subplot) and Bregman distances $D_{j_q}(x^\delta_{N(\delta, y^\delta)}, x^\dagger)$ (dots, right subplot) as functions of the noise level δ. The solid line in the small subplot has the slope corresponding to the rate predicted by the theory. (x-axis: noise level δ; y-axis: stopping index $N(\delta, y^\delta)$ resp. Bregman distance $D_{j_q}(x^\delta_{N(\delta, y^\delta)}, x^\dagger)$)

7

Final Remarks

In this chapter, we present final remarks concerning different topics discussed in this thesis. We conclude this chapter with the list of new contributions presented in this thesis. We begin with various remarks:

1. (Geometry of Besov spaces) It seems to be a common knowledge that there exists a renorming of Besov spaces $B_{p,q}^s(\mathbb{R})$, such that the resulting norm is $\min\{2, p, q\}$-smooth and $\min\{2, p, q\}$-convex, cf. [12]. However, to the best of author's knowledge, it has never been proven that the wavelet characterization of the Besov norm is an example of such a renorming. This was the motivation for the results presented in Chapter 3.

2. (Variational source conditions) The authors of [35] and [55, Section 3.2] introduced a new concept of the so-called *variational source condition*, which is given by

$$\langle j_q(x^\dagger), x^\dagger - x \rangle \leq \gamma \cdot D_{j_q}(x^\dagger, x) + \Gamma \cdot \|Ax^\dagger - Ax\| \qquad (7.1)$$

for some $\gamma < 1$ and $\Gamma \geq 0$ and all $x \in X$. For non-linear operators the variational source condition is a generalization of source conditions previously introduced in the literature for non-linear operators. On the contrary, it was shown in [25, Proposition 12 and 13] that for linear operators the low order source condition $j_q(x^\dagger) = A^*\omega$ is equivalent to the variational source condition (7.1). However, with the variational source condition we have a tool to weaken the low order source condition. For example in [30] the authors propose the source condition

$$\langle j_q(x^\dagger), x^\dagger - x \rangle \leq \gamma \cdot D_{j_q}(x^\dagger, x) + \Gamma \cdot \frac{1}{\kappa}\|Ax - Ax^\dagger\|^\kappa,$$

where $\gamma < 1, \Gamma > 0$ and

$$0 < \kappa \leq 1.$$

To the best of author's knowledge no characterization of the above weakened variational source condition in terms of range source conditions is known.

3. (Approximate source conditions) In some problems the low order source condition $j_q(x^\dagger) = A^*\omega$ is not valid but we know that the range of A^* is dense in X^*. Hence, for every $\varepsilon > 0$ there exist elements ω and v with $\|v\| \leq \varepsilon$ such that

$$\xi = A^*\omega + v.$$

Therefore, the low order source condition holds approximately. This idea is used in the construction of the so called *approximate source conditions* and was introduced in [5]. For every $S \geq 0$ we define the function d via

$$d(S) := \inf\{\|x^\dagger - A^*\omega\| \ : \ \|\omega\| \leq S\}.$$

Then, there exist $\omega(S)$ and $v(S)$ such that

$$x^\dagger = A^*\omega(S) + v(S) \qquad \text{and} \qquad \|\omega(S)\| \leq S, \|v(S)\| = d(S).$$

The convergence rates for parameter choice rules are then described via decay properties of the function d. The results of e.g. [33, 34, 30, 27, 26] show that this idea turns out to be very powerful.

4. (General setting for source conditions) Despite the above generalizations, the regularization theory in Banach spaces still lacks an unifying approach, which would generalize the low order and the high order source conditions in a natural and intuitive way.

5. (Optimality results) To the best of author's knowledge, it is still unclear whether the convergence rates for the Bregman distance $D_{j_q}(x^\dagger, x_\alpha^\delta)$ discussed in Chapter 4 are optimal. First steps in this direction were made in [52].

6. (Sensitivity analysis) In this thesis the penalty term in the Tikhonov functional was fixed. An interesting question is how the choice of the penalty term influences the minimizers of the Tikhonov functional, e.g. how much do the minimizers x_{α,ℓ^p}^δ for $p > 1$ resemble x_{α,ℓ^1}^δ?

7. (Steepest descent in the primal space and dual space for sums of convex functionals) The Tikhonov functional $\mathrm{T}_\alpha(x) := \frac{1}{p}\|Ax - y^\delta\|^p + \alpha \cdot \frac{1}{q}\|x\|^q$ is a special case of the functional

$$\Psi(x) := F(x) + \Phi(x),$$

where F and Φ are convex functionals. The steepest descent in the primal space

$$x_{n+1} = x_n - \mu_n J_{q^*}(\nabla \mathrm{T}_\alpha(x_n))$$

can then be generalized for Ψ by

$$x_{n+1} = x_n - \mu_n J_{q*}(\nabla \Psi(x_n))$$

Moreover, steepest descent in the dual space

$$x_{n+1} = J_{q*}(J_q(x_n) - \mu_n \partial \Psi(x_n))$$

may be generalized via

$$x_{n+1} = (\partial \Phi)^{-1}(\partial \Phi(x_n) - \mu_n \partial \Psi(x_n)).$$

Convergence proofs for the above generalizations are subject of ongoing research.

8. (Connections to accretive operators) We remark that the methods used in the proof of convergence rates for the steepest descent in the dual space may also be used in the study of properties of Mann iterations for accretive operators. This idea was published by the author in [41] and [42].

9. (Other iterative regularization methods) Recently other types of iterative regularization in Banach spaces have been considered. In [60], the authors discuss a method based on subspace optimization, whereas in [40] a variant of the iteratively regularized Gauss-Newton method is analyzed.

10. (High order source conditions for Landweber iteration) An interesting open problem is to determine whether it is possible to show high order convergence rates for the Landweber iteration under high order source condition $j_q(x^\dagger) = A^* j_p(A\omega)$.

We conclude this thesis with a list of new contributions:

1. In Chapter 3, we showed that the characterization of the norm of Besov spaces $B_{p,q}^s(\mathbb{R})$ is convex of power type and smooth of power type for $1 < p, q < \infty$. To the best of author's knowledge this result is new.

2. In Section 4.3.3, we developed a generalization of the discrepancy principle of Engl to the setting of Banach spaces. We showed regularization properties and convergence rates. To the best of author's knowledge there exists only one other method of an a-posteriori parameter choice rule for which high order convergence rates have been shown, cf. [28]. The results of this thesis give an alternative approach to this problem.

3. The convergence rates for the steepest descent in the primal space proven in Section 4.4.1 are new. We note that the proof of strong convergence, but without particular convergence rates, was already published in [7].

4. In Section 4.4.2, we proved convergence rates for the steepest descent in the dual space. A slightly different version of the proof was published by the author in [43].

5. We remark that the ideas for accelerating the Landweber iteration of [59] can already be found in [58] and [60]. However, the main problem of the approaches discussed there is that only weak convergence can be shown. Therefore, the Section 5.1 contains two new contributions. On the one hand, we developed a version of the Landweber iteration, which is adapted to Banach spaces convex of power type and smooth of power type. On the other hand, we were able to overcome the aforementioned problems and proved strong convergence for the adapted, accelerated version of Landweber iteration. Based on the ideas developed in this thesis, Hein generalized the accelerated Landweber iteration to a class of nonlinear operators and published these results together with the author in [31].

6. The main contribution of Section 5.2 is the introduction of modified Landweber iteration and the proof of convergence rate with respect to the Bregman distance under the low rate source condition. Based on the ideas developed in this thesis, Hein generalized the modified Landweber iteration to a class of nonlinear operators and to a more general class of source conditions and published these results together with the author in [31].

7. The last contribution is the idea of solving the problem of Engl's discrepancy principle with model functions, as presented in Section 6.3. Further research is necessary in order to prove the convergence of this method.

Proof of regularization properties of Tikhonov functionals

In this appendix we will prove Theorem 4.2. We start with the the following general result:

Theorem A.1. *[55, Section 3.2] Let X, Y be Banach spaces and $F : \mathcal{D}(F) \subset X \to Y$ be a (non-)linear operator. Let the Tikhonov functional be given by*

$$\mathrm{T}_\alpha(x) := \tfrac{1}{p}\|F(x) - y^\delta\|_Y^p + \alpha\mathcal{P}(x),$$

where

1. *the Banach spaces X and Y are associated with topologies τ_X and τ_Y which are weaker than the norm topologies,*

2. *the exponent p is greater than or equal 1,*

3. *the norm $\|\cdot\|_Y$ is sequentially lower semi-continuous with respect to τ_Y,*

4. *the functional $\mathcal{P} : X \to [0, \infty]$ is convex and sequentially lower semi-continuous with respect to τ_X, i.e. if for all sequences (x_n) converging to x with respect to the τ_X-topology we have*

$$\liminf_n \mathcal{P}(x_n) \geq \mathcal{P}(x),$$

5. *the set $\mathcal{D} := \mathcal{D}(F) \cap \mathrm{dom}(\mathcal{P})$ is not empty,*

6. *for every $\alpha > 0$ and $M > 0$, the level sets*

$$\mathcal{M}_\alpha(M) := \{x \in X \ : \ \mathrm{T}_\alpha(x) \leq M\}$$

are sequentially pre-compact with respect to τ_X, i.e. every sequence in $\mathcal{M}_\alpha(M)$ has a sub-sequence which converges with respect to the τ_X-topology,

7. for every $\alpha > 0$ and $M > 0$, the set $\mathcal{M}_\alpha(M)$ is sequentially closed with respect to τ_X and the restriction of F to $\mathcal{M}_\alpha(M)$ is sequentially continuous with respect to the topologies τ_X and τ_Y,

8. the problem $F(x) = y$ has a solution in $\mathcal{D}$.

Then

1. (Existence) For every $\alpha > 0$ and $y^\delta \in Y$ there exists a minimizer of T_α.

2. (Stability) If (y_k) is a sequence converging to y^δ in Y with respect to the norm topology, then every sequence (x_k) with

$$x_k \in \mathrm{argmin}_{x \in \mathcal{D}(F)} \tfrac{1}{p} \|F(x) - y_k\|_Y^p + \alpha \mathcal{P}(x),$$

has a sub-sequence that converges with respect to τ_X. The limit of every τ_X-convergent sub-sequence $(x_{k'})$ of (x_k) is a minimizer z of $\tfrac{1}{p}\|F(x) - y^\delta\|_Y^p + \alpha \mathcal{P}(x)$ and the sequence $(\mathcal{P}(x_{k'}))$ converges to $\mathcal{P}(z)$.

3. There exists a $\mathcal{P}$-minimizing solution $x^\dagger$ of $F(x) = y$, where $x^\dagger$ is an $\mathcal{P}$-minimizing solution if

$$F(x^\dagger) = y \quad \text{and} \quad \mathcal{P}(x^\dagger) = \inf\{\mathcal{P}(x) \ : \ x \in \mathcal{D}(F), F(x) = y\}.$$

4. (Convergence) Assuming $\alpha(\delta, y^\delta)$ satisfies for all y^δ with $\|y - y^\delta\| \le \delta$

$$\alpha(\delta, y^\delta) \to 0 \quad \text{and} \quad \frac{\delta^p}{\alpha(\delta, y^\delta)} \to 0, \quad \text{as} \quad \delta \to 0$$

moreover, assuming that the sequence (δ_k) converges to 0, and that y^{δ_k} satisfies $\|y - y^{\delta_k}\| \le \delta_k$. Then every sequence (x_k) of elements minimizing $\tfrac{1}{p}\|F(x) - y^{\delta_k}\|_Y^p + \alpha(\delta_k, y_k) \cdot \mathcal{P}(x)$ has a sub-sequence $(x_{k'})$ that converges with respect to τ_X. The limit $x^\dagger$ of every τ_X-convergent sub-sequence $(x_{k'})$ is an $\mathcal{P}$-minimizing solution of $F(x) = y$ and $\mathcal{P}(x_k) \to \mathcal{P}(x^\dagger)$.

If in addition to the assumptions above:

1. τ_X is the weak topology on X,

2. the functional $\mathcal{P}$ is totally convex, i.e. every sequence (x_k) in $\mathrm{dom}\,\mathcal{P}$ with

$$(\mathcal{P}(x_k) - \mathcal{P}(x) - \langle \partial \mathcal{P}(x), x_k - x \rangle) \to 0$$

for some $x \in X$ satisfies $\|x_k - x\| \to 0$ and

3. the $\mathcal{P}$-minimizing solution $x^\dagger$ is unique.

Then:

1. (Strong convergence) Assuming $\alpha(\delta, y^\delta)$ satisfies for all y^δ with $\|y - y^\delta\| \leq \delta$

$$\alpha(\delta, y^\delta) \to 0 \quad \text{and} \quad \frac{\delta^p}{\alpha(\delta, y^\delta)} \to 0, \quad \text{as} \quad \delta \to 0$$

moreover, assuming that the sequence (δ_k) converges to 0, and that y^{δ_k} satisfies $\|y - y^{\delta_k}\| \leq \delta_k$. Then, for every sequence (x_k) of elements minimizing $\frac{1}{p}\|F(x) - y^{\delta_k}\|_Y^p + \alpha(\delta_k, y_k) \cdot \mathcal{P}(x)$ we have

$$x_k \to x^\dagger$$

with respect to the norm topology.

Remark A.2. Let X be a reflexive Banach space, Y an arbitrary Banach space, $A : X \to Y$ a linear and continuous operator and T_α the Tikhonov functional defined via

$$\mathrm{T}_\alpha(x) = \tfrac{1}{p}\|Ax - y^\delta\|_Y^p + \alpha \cdot \tfrac{1}{q}\|x\|_X^q \qquad p, q > 1$$

with $\|y^\delta - y\| \leq \delta$ and $Ax = y$ for some x. Further, let τ_X and τ_Y be chosen as the weak topologies on X and Y, where (x_n) is said to be weakly convergent to x if $\langle x^*, x_n \rangle \to \langle x^*, x \rangle$ for all elements x^* of the dual space. Then, the first part of the assumptions of the last theorem is fulfilled, since

1. the weak topology is weaker than the strong topology,

2. by assumption $p > 1$,

3. the norm $\|\cdot\|_Y$ is sequentially lower semi-continuous with respect to the weak topology (cf. [51, Theorem 2.5.21]),

4. the functional $\frac{1}{q}\|\cdot\|_X^q$ is sequentially lower semi-continuous with respect to the weak topology since the norm $\|\cdot\|_X$ is sequentially lower semi-continuous with respect to the weak topology (cf. [51, Theorem 2.5.21]) and $x \to \frac{1}{q}x^q$ is monotonically increasing and continuous on the range of $\|\cdot\|_X$, i.e. $[0, \infty)$,

5. we have $\mathcal{D} = \mathcal{D}(A) \cap \mathrm{dom}(\frac{1}{q}\|\cdot\|_X^q) = X \cap X = X \neq \emptyset$,

6. every sequence in the the level set

$$\mathcal{M}_\alpha(M) := \{x \in X \; : \; \tfrac{1}{p}\|Ax - y\|_Y^p + \alpha \cdot \tfrac{1}{q}\|x\|_X^q \leq M\}$$

is bounded due to

$$\alpha \cdot \tfrac{1}{q}\|x\|_X^q \leq \tfrac{1}{p}\|Ax - y\|_Y^p + \alpha \cdot \tfrac{1}{q}\|x\|_X^q \leq M$$

and therefore has a weakly convergent sub-sequence due to reflexivity of X (c.f. [51, Theorem 2.8.9]) and therefore $\mathcal{M}_\alpha(M)$ is weakly pre-compact,

7. every set $\mathcal{M}_\alpha(M)$ is weakly sequentially closed, since the the Tikhonov functional T_α is sequential lower semi-continuous, hence weakly lower semi-continuous (cf. [21, Corollary 2.2]), hence every set $\mathcal{M}_\alpha(M)$ is weakly sequentially closed (cf. [36, Theorem 2.5]); further the operator A is weak-to-weak continuous (cf. [51, Theorem 2.5.11]),

8. and by assumption $Ax = y$ has a solution in $\mathcal{D} = X$.

Proof. (of Theorem 4.2) By assumption the space X is convex of power type. Therefore, by Theorem 2.41, X is reflexive and the claims of Remark A.2 hold with τ_X and τ_Y being the weak topologies. The second part of assumptions of Theorem A.1 holds, since

1. τ_X is the weak topology,

2. The functional $\frac{1}{q}\|\cdot\|^q$ is convex of power type and therefore totally convex by Theorem 2.48.

3. the norm minimizing solution $x^\dagger$ is unique by Lemma 2.55.

□

Bibliography

[1] R. A. Adams. *Sobolev Spaces*. Academic Press, New York, 1975.

[2] Y. I. Alber, A. N. Iusem, and M. V. Solodov. Minimization of nonsmooth convex functionals in Banach spaces. *Journal of Convex Analysis*, 4(2):235–255, 1997.

[3] E. Asplund. Positivity of duality mappings. *Bulletin of the American Mathematical Society*, 73(2):200–203, 1967.

[4] A. B. Bakushinskii. Remarks on choosing a regularization parameter using the quasi-optimality and ratio criterion. *USSR Computational Mathematics and Mathematical Physics*, 24(4):181–182, 1984.

[5] J. Baumeister. *Stable Solutions of Inverse Problems*. Vieweg, Braunschweig, 1987.

[6] T. Bonesky. Morozov's discrepancy principle and Tikhonov-type functionals. *Inverse Problems*, 25(1):015015 (11pp), 2009.

[7] T. Bonesky, K. S. Kazimierski, P. Maaß, F. Schöpfer, and T. Schuster. Minimization of Tikhonov functionals in Banach spaces. *Abstract and Applied Analysis*, Volume 2007:192679 (19 pp), 2007.

[8] K. Bredies and D. A. Lorenz. Iterated hard shrinkage for minimization problems with sparsity constraints. *SIAM Journal on Scientific Computing*, 30(2):657–683, 2008.

[9] L. M. Bregman. The relaxation method for finding common points of convex sets and its application to the solution of problems in convex programming. *USSR Computational Mathematics and Mathematical Physics*, 7(3):200–217, 1967.

[10] M. Burger and S. Osher. Convergence rates of convex variational regularization. *Inverse Problems*, 20(5):1411–1421, 2004.

[11] I. Cioranescu. *Geometry of Banach spaces, duality mappings and nonlinear problems*. Kluwer, Dordrecht, 1990.

[12] F. Cobos and D. E. Edmunds. Clarkson's inequalities, Besov spaces and Triebel-Sobolev spaces. *Zeitschrift für Analysis und ihre Anwendungen*, 7(3):229–232, 1988.

[13] C. Cohen. *Numerical analysis of wavelet methods*, volume 32 of *Studies in Mathematics and its Applications*. North-Holland, Amsterdam, 2003.

[14] I. Daubechies. *Ten Lectures on Wavelets*. SIAM, New York, 1992.

[15] I. Daubechies, M. Defrise, and C. De Mol. An iterative thresholding algorithm for linear inverse problems with a sparsity constraint. *Communications in Pure and Applied Mathematics*, 57(11):1413–1457, 2004.

[16] R. Deville, G. Godefroy, and V. Zizler. *Smoothness and Renormings in Banach spaces*. Pitman Monographs and Surveys in Pure and Applied Mathematics. Longman Scientific & Technical, Harlow, UK, 1993.

[17] R. A. DeVore. Nonlinear approximation. *Acta Numerica*, 7:51–150, 1998.

[18] N. Dunford and J. T. Schwartz. *Linear Operators Part I: General Theory*. John Wiley and Sons, New York, 1988.

[19] J. C. Dunn. Rates of convergence for conditional gradient algorithms near singular and nonsingular extremals. *SIAM Journal on Control and Optimization*, 17(2):187–211, 1979.

[20] J. C. Dunn. Global and asymptotic convergence rate estimates for a class of projected gradient processes. *SIAM Journal on Control and Optimization*, 19(3):368–400, 1981.

[21] I. Ekeland and R. Temam. *Convex analysis and variational problems*. North-Holland, Amsterdam, 1976.

[22] H. W. Engl. Discrepancy principles for Tikhonov regularization of ill-posed problems leading to optimal convergence rates. *Journal of Optimization Theory and Applications*, 52(2):209–215, 1987.

[23] H. W. Engl, M. Hanke, and A. Neubauer. *Regularization of Inverse Problems*, volume 375 of *Mathematics and its Applications*. Kluwer Academic Publishers Group, Dordrecht, 2000.

[24] M. Frazier, B. Jawerth, and G. Weiss. *Littlewood-Paley theory and the study of function spaces*. Number 79 in Regional Conference Series in Mathematics. American Mathematical Society, Boston, 1999.

[25] M. Grasmair, M. Haltmeier, and O. Scherzer. Sparse regularization with ℓ^q penalty term. *Inverse Problems*, 24(3):055020 (13pp), 2008.

[26] T. Hein. Convergence rates for regularization of ill-posed problems in Banach spaces by approxinative source coniditions. *Inverse Problems*, 24(4):045007 (10pp), 2008.

[27] T. Hein. Regularization in Banach spaces — convergence rates by approximative source conditions. *Journal of Inverse and Ill-Posed Problems*, 17:27–41, 2009.

[28] T. Hein. *Regularization in Banach spaces — convergence rates theory*. Habilitation thesis, Technische Universität Chemnitz, Chemnitz, 2009.

[29] T. Hein. Tikhonov regularization in Banach spaces — improved convergence rates results. *Inverse Problems*, 25(3):035002 (18pp), 2009.

[30] T. Hein and B. Hofmann. Approximate source conditions for nonlinear ill-posed problems — chances and limitations. *Inverse Problems*, 25(3):035003 (16pp), 2009.

[31] T. Hein and K. S. Kazimierski. Accelerated Landweber iteration in Banach spaces. *Inverse Problems*, 26(5):055002 (17pp), 2008.

[32] T. Hein and K. S. Kazimierski. Modified Landweber iteration in Banach spaces — convergence and convergence rates. Preprint 14, Technische Universität Chemnitz, Faculty of Mathematics, 2009.

[33] B. Hofmann. Approximate source conditions in Tikhonov-Phillips regularization and consequences for inverse problems with multiplication operators. *Mathematical Methods in the Applied Sciences*, 29(3):351–371, 2006.

[34] B. Hofmann, D. Düvelmeyer, and K. Krumbiegel. Approximate source conditions in regularization — new analytical results and numerical studies. *Journal Mathematical Modelling and Analysis*, 11(1):41–56, 2006.

[35] B. Hofmann, B. Kaltenbacher, C. Pöschl, and O. Scherzer. A convergence rates result for Tikhonov regularization in Banach spaces with non-smooth operators. *Inverse Problems*, 23(3):987–1010, 2007.

[36] J. Jahn. *Introduction to the theory of nonlinear optimization*. Springer, Berlin, 2007.

[37] B. Jin and J. Zou. Iterative choice of regularization parameter via discrepancy principle. *in preparation*, 2009.

[38] Q.-N. Jin and Z.-Y. Hou. On the choice of the regularization parameter for ordinary and iterated Tikhonov regularization of nonlinear ill-posed problems. *Inverse Problems*, 13(3):815–827, 1997.

[39] G. Kaiser. *A Friendly Guide to Wavelets*. Birkhauser, Boston, 1999.

[40] B. Kaltenbacher, F. Schöpfer, and T. Schuster. Iterative methods for nonlinear ill-posed problems in Banach spaces: convergence and applications to parameter identification problems. *Inverse Problems*, 25(6):065003 (19pp), 2009.

[41] K. S. Kazimierski. Adaptive Mann iterations for nonlinear accretive and pseudocontractive operator equations. *Mathematical Communications*, 13(1):33–44, 2008.

[42] K. S. Kazimierski. Are adaptive Mann iterations really adaptive? *Mathematical Communications*, 14(2):399–412, 2009.

[43] K. S. Kazimierski. Minimization of the Tikhonov functional in Banach spaces smooth and convex of power type by steepest descent in the dual. *to appear in Computational Optimization and Applications*, 2009.

[44] P. Kügler. Convergence rate analysis of a derivative free Landweber iteration for parameter identification in certain elliptic PDEs. *Numerische Mathematik*, 101(1):165–184, 2009.

[45] J. Lindenstrauß and L. Tzafriri. *Classical Banach Spaces, II*. Springer, Berlin, 1979.

[46] D. Lorenz. *Wavelet Shrinkage in Signal and Image Processing - An Investigation of Relations and Equivalences*. PhD thesis, Universität Bremen, Bremen, 2005.

[47] A. K. Louis, P. Maaß, and A. Rieder. *Wavelets: Theory and Applications*. Wiley, New York, 1997.

[48] S. Mallat. *A Wavelet Tour of Signal Processing*. Academic Press, New York, 1999.

[49] V. Maz'ya. *Sobolev Spaces in Mathematics I: Sobolev Type Inequalities*. Springer, Berlin, 2008.

[50] V. Maz'ya. *Sobolev Spaces in Mathematics II: Applications in Analysis and Partial Differential Equations*. Springer, Berlin, 2008.

[51] R. E. Megginson. *An Introduction to Banach Space Theory*. Springer, New York, 1998.

[52] A. Neubauer. On enhanced convergence rates for Tikhonov regularization of nonlinear ill-posed problems in Banach spaces. *Inverse Problems*, 25(6):065009 (10pp), 2009.

[53] E. Resmerita. Regularization of ill-posed problems in Banach spaces: convergence rates. *Inverse Problems*, 21(4):1303–1314, 2005.

[54] R. T. Rockafellar and R. J.-B. Wets. *Variational Analysis*. Springer, Berlin, 1998.

[55] O. Scherzer, M. Grasmair, H. Grossauer, M. Haltmeier, and F. Lenzen. *Variational Methods in Imaging*. Springer, New York, 2008.

[56] W. Schirotzek. *Nonsmooth Analysis*. Springer, Berlin, 2007.

[57] E. Schock. Approximate solution of ill-posed equations: Arbitrarily slow convergence vs. superconvergence. In G. Hämmerlin and K. Hofmann, editors, *Constructive Methods for the Practical Treatment of Integral Equations*, pages 234–243. Birkhäuser, Basel, 1985.

[58] F. Schöpfer. *Iterative Regularization Methods for the Solution of the Split Feasibility Problem in Banach Spaces*. PhD thesis, Universität des Saarlandes, Saarbrücken, 2007.

[59] F. Schöpfer, A. K. Louis, and T. Schuster. Nonlinear iterative methods for linear ill-posed problems in Banach spaces. *Inverse Problems*, 22(1):311–329, 2006.

[60] F. Schöpfer and T. Schuster. Fast regularizing sequential subspace optimization in Banach spaces. *Inverse Problems*, 25(1):015013 (22pp), 2009.

[61] F. Schöpfer, T. Schuster, and A. K. Louis. Metric and Bregman projections onto affine subspaces and their computation via sequential subspace optimization methods. *Journal of Inverse and Ill-Posed Problems*, 16(5):479–506, 2008.

[62] R. E. Showalter. *Monotone operators in Banach spaces and nonlinear partial differential equations*. American Mathematical Society, Providence, 1997.

[63] S. L. Sobolev. The Cauchy problem in a function space (in Russian). *Doklady Akademii Nauk SSSR*, 3(7):291–294, 1935.

[64] S. L. Sobolev. Méthode nouvelle à resoudre le problème de Cauchy pour les équations linéaires hyperboliques normales. *Matematicheskii Sbornik*, 1(1):39–72, 1936.

[65] S. L. Sobolev. On a theorem of functional analysis (in Russian). *Matematicheskii Sbornik*, 4(3):471–497, 1938.

[66] H. Triebel. *Theory of function spaces*. Birkhäuser, Basel, 1983.

[67] H. Triebel. *Theory of function spaces II*. Birkhäuser, Basel, 1992.

[68] H. Triebel. *Theory of function spaces III*. Birkhäuser, Basel, 2006.

[69] J. Xie and J. Zou. An improved model function method for choosing regularization parameters in linear inverse problems. *Inverse Problems*, 18(3):631–643, 2002.

[70] Z.-B. Xu. Characteristic inequalites of l^p spaces and their applications (in Chinese). *Acta Mathematica Sinica (Chinese series)*, 32(2):209–218, 1989.

[71] Z.-B. Xu and G. Roach. Characteristic inequalities of uniformly convex and uniformly smooth Banach spaces. *Journal of Mathematical Analysis and Applications*, 157(1):189–210, 1991.

[72] Z.-B. Xu and Z.-S. Zhang. Another set of characteristic inequalities of l^p Banach spaces (in Chinese). *Acta Mathematica Sinica (Chinese series)*, 37(4):433–439, 1994.

[73] E. Zeidler. *Nonlinear Functional Analysis and its Applications III*. Springer-Verlag, New York, 1984.